CW00530260

Analogies in Physics and Life

A Scientific Autobiography

Analogies in Physics and Life

A Scientific Autobiography

Richard M Weiner

Université Paris-Sud, France
University of Marburg, Germany

World Scientific

NEW JERSEY • LONDON • SINGAPORE • BEIJING • SHANGHAI • HONG KONG • TAIPEI • CHENNAI

Published by

World Scientific Publishing Co. Pte. Ltd.
5 Toh Tuck Link, Singapore 596224
USA office: 27 Warren Street, Suite 401-402, Hackensack, NJ 07601
UK office: 57 Shelton Street, Covent Garden, London WC2H 9HE

British Library Cataloguing-in-Publication Data
A catalogue record for this book is available from the British Library.

ANALOGIES IN PHYSICS AND LIFE
A Scientific Autobiography

ISBN-13 978-981-270-470-2
ISBN-10 981-270-470-1
ISBN-13 978-981-270-471-9 (pbk)
ISBN-10 981-270-471-X (pbk)

Typeset by Stallion Press
Email: enquiries@stallionpress.com

Printed in Singapore by Mainland Press Pte Ltd

To the memory of my parents

Contents

Preface

Analogies in science are used either to explain certain observed experimental facts or to predict new effects. An appreciable fraction, if not the majority, of important discoveries in modern physics is based on analogies, and one could take the point of view that the entire evolution of science is determined by this way of thinking, even if those who were involved were not always conscious of this.

Despite the importance of this fundamental fact, there is apparently no book in the scientific literature (at least in the physical sciences) dedicated to this subject. Moreover, to understand how and why at a certain moment a certain analogy was used one has to know the specific historical circumstances under which the new idea was developed. While a particular analogy itself may be explained in scientific journals or in textbooks, the historical background is usually not. These historical circumstances are of interest and often essential—for the undergraduate or graduate student who learns the subject for the first time, but also for the practitioner who looks for inspiration or who wants to understand what his colleagues in other fields did and why, and last but not least for the historian of science.

The present book is an attempt to contribute to filling this gap, albeit a quite subjective one: I tried to address this issue by referring to different phenomena in subatomic physics studied or predicted over the last decades and to which I contributed personally. Some of the predictions have been confirmed, others are still in the pipeline and quite a few might never be verified or even proven wrong. Still, I hope that even those in the last category might be useful, if not for any other reason than to prevent somebody else from repeating these mistakes. Given the general readership to which this book addresses itself, I attempted to present the effects discussed at a level accessible to the non-specialist.

Although the notion of analogy originated in the exact sciences, and more precisely in mathematics (in Greek *ana logon*, means "according to a ratio" and referred initially to a *similarity* between two figures), it was soon extended to other domains of science, including history. This is reflected also in the content of this book when discussing the personal and historical background. Indeed, it is this background which almost by itself led to analogies in life as well: I had the "privilege" of living under two dictatorships, the Nazi and the Communist ones, and I was lucky enough to survive, while a large part of my family did not. This motivated me to try to draw parallels between the two regimes. Furthermore, in the second part of my life, I was fortunate to live in different democratic countries, most recently in Germany and France. This stimulated my attempt at comparing life in these two countries, again, of course, from quite a personal point of view.

I am indebted to Apy Vourdas and Uli Weiner for carefully reading the manuscript and for their valuable comments.

PART I

The Wandering Years (1930–1974)

Section I

Czernowitz, a City of People and Books that No Longer Exists (1930–1945)

1

Childhood

My Countries

When I am asked what country I was born in, I often hesitate before answering: Am I supposed to name the country to which the city in which I was born belonged at the time I was born, or the country it belonged to when I left it? Or, perhaps, the country to which it belongs now? Or rather the country it belonged to just 12 years before I was born, the country where my parents and grandparents were born, the country of my mother tongue?

My problem is that these were four different countries; only within the fifteen years I lived there, the place where I was born changed "ownership" three times. This frequent, and for most of its citizens, tragic change is reflected in the following joke. Two people meet for the first time. After a few minutes of conversation they ask each other where they are from. It turns out that one of them is from Czernowitz. The other, who had never heard about Czernowitz asks: "Czernowitz, where is this?" "Sorry," is the answer, "I am not sure, I haven't read today's newspaper yet."

I happen to be born in that Czernowitz, the capital of Bukowina, a province situated in the Eastern part of Central Europe. The population of the Bukowina was of mixed origin: in the rural regions lived Romanians, Ukrainians (in Austria-Hungary they were called Ruthenians), and in lesser proportions Jews, Germans and Poles. In the cities, and in Czernowitz in particular, the situation was quite different: In 1940 Czernowitz had a population of about 150,000 people, out of which about 60,000 were Jews. Here a parenthesis has to be made: When referring to Romanians, Ukrainians, Germans and Poles *origin* means nationality. For Jews this word has in the present context a different meaning, it means essentially religion.[a] This is due to the fact that until the end of World War I, with the notable exception of Austria-Hungary, in Central and Eastern Europe Jews had no civil rights. Romania had in this respect a particularly (in)famous record. It was the last European country to grant in 1923, and that only due to the pressure of the West, Jews citizenship.

During the period 1775–1918, Bukowina was part of the Austro-Hungarian Empire and my parents and grandparents were Austrian citizens. Strangely enough, although I was born in 1930, when Czernowitz and the entire Bukowina belonged to Romania, and although I spent, over all, 37 years in Romania, I have always considered myself Austrian rather than Romanian. This was a fortiori true for my parents and grandparents. And now, as a holder of a German passport, I am happy to be able to consider myself first of all a citizen of Europe.

Most of the population of Bukowina, at least the German speaking part, felt like us. And German was *the* language of Bukowina and of Czernowitz in particular, because most people in Bukowina, even after 1918 when they became Romanian, spoke German. This period came to an end with World War II. Though, it might be interesting to mention that a few years ago German newspapers[1] and TV[2] discovered three survivors of this Austrian/German culture island, who still lived in Czernowitz. One of these three survivors, Mathias Zwilling, who died in 1999,

[a]To understand better this etymological difference it should be mentioned that Romanians and Ukrainians were in general Orthodox Christians, and Germans and Poles mostly Catholics.

happened to have been a youth friend and elementary school classmate of mine. Two years before his death I managed to get in contact with him with the help of the journalist Meyer-Timpe from *Die Zeit* who had interviewed him.

Bukowina has had a particular status within the Austrian-Hungarian Empire. It was a "crown-land", which meant that it was directly ruled by Vienna. Unlike Tsarist Russia or Romania, in Austria the population of Jewish origin, like my ancestors, had full citizen rights. Jews occupied high-ranking positions in administration and politics. Czernowitz had Jewish mayors; Bukowinian Jews were deputies in the Vienna Federal Parliament, professors at the Kaiser Franz Joseph University of Czernowitz and at other Austrian universities, including that of Vienna, and played in Bukowina a major, if not dominating role, in commerce, liberal professions, and first of all in culture. And there was quite an active cultural life in Bukowina, not only before 1918, but also after that, until it took an abrupt end in the early 1940s with the physical extermination of most of the Jewish population and with the Diaspora of the few who survived. The man who is considered today as the greatest German post-war poet–according to *The New York Times*[3] even the greatest post-war European poet, Paul Celan, was born in Czernowitz and spent the first 25 years of his life there, before emigrating via Romania to France. He survived the ghetto only because, due to a curfew, he did not sleep at home the night preceding the deportation of his parents. The poet Selma Meerbaum-Eisinger, with whom I was remotely related, perished in the camps of Transnistria at the age of 18. She continued to write even in the camp. Joseph Schmidt, born in (the neighborhood of) Czernowitz also found a tragic end during the war. In 1942 this famous tenor, dubbed the German Caruso, was fleeing from the Nazis. He asked for and was refused asylum in Switzerland. He then crossed illegally the French-Swiss border and was interned in a camp, where he died, at the age of 38, because of an inadequately treated cold. His fate had impressed me in particular, not only because I liked his voice very much, but also because he was a cousin of a Czernowitz schoolmate of mine, Ari Rosenbach, who used to tell me in the 1930s about him and his musical career. I met Ari again, incidentally, after about 40 years, in Sao Paulo. Here I also met Martha Radsprecher, another former Meisler schoolmate of mine. Both Ari and Martha had managed to escape from Romania shortly after 1945. Two other Meisler schoolmates I met after war are Margot Gottesmann-Ringwald who lives now in Basel and Friedel Sonntag, who lives in Tel Aviv. Friedel managed to escape from Romania in 1951 already and has been the first PhD of the Weizmann Institute.

The Diaspora list of German language poets and writers coming from the Bukowina is much longer. All of them spent the rest of their life abroad. To mention just a few names and the countries they went to: Moses Rosenkranz — Germany, Gregor von Rezzori (who wasn't Jewish) — Germany and Italy, Alfred Gong — USA, Klara Blum — China, Manfred Winkler — Israel, Immanuel Weißglas — Romania, Alfred Margul-Sperber — Romania (he had been a friend of my father), Rose Ausländer — Germany, Alfred Kittner — Germany, Ninon Ausländer-Hesse (the wife of Hermann Hesse) — Switzerland. Last but not least, I should mention also that the famous scientist Erwin Chargaff, the discoverer of the DNA formula, was born in 1905 in Czernowitz. He worked and spent most of his life in the USA, where he died in 2002.

My Languages

Although in 1918 Romanian was spoken by part of the rural population, for the rest of the population of Bukowina it was a foreign language. An appreciable proportion of the rural population spoke Ruthenian, while in Czernowitz and most other smaller cities (like Storojinets, where my father's parents, brothers and sisters lived) most people spoke German. Until the end of the First World War in 1918, German had been both the official language and the language of culture, and it maintained its dominating role, both in commerce and in culture, also under Romanian rule, in the period 1918–1940.

In Czernowitz appeared five daily German newspapers and in the cafés one could find every day not less than 150 newspapers, mostly in German, from all over the world. In the libraries and bookshops German books dominated. Some of the newspapers published in Czernowitz circulated all over the Europe, and in particular in Austria and Germany, even after World War I, when Bukowina was attributed to Romania. The following quotation from Friedrich Thorberg, one of the most important Austrian writers of the 20th century, is quite significant from this point of view: "The real and definitive end of the Austrian-Hungarian Empire happened with Hitler's entry in Vienna, that is twenty years after its political dissolution in November 1918. The curtain was drawn only when the *Prager Tageblatt* and the *Czernowitzer Morgenzeitung* were not anymore on the tables of the Viennese cafes." That explains why, although I was born in 1930, when Czernowitz belonged to Romania, my native language is German. In the elementary school I attended in the period 1936–1940, at the Meisler School, German was taught from the very first year.

I mentioned the cafés. Like its big brother, Vienna, the pre- war Czernowitz I remember was a city of *Kaffeehäuser*. The Austrian cafés were a kind of extension of offices and flats, where people spent an appreciable part of the day. My father and the brothers of my mother frequented them quite often; my father used to play chess in cafe *Bellevue* situated on the *Ringplatz*, between *Rathausstrasse* and *Herrengasse,* and I sometimes accompanied him to kibitz. I found this café reproduced on a 1905 photograph of Czernowitz in the book by Florence Heymann: "Czernowitz, Etincelles de Mémoires," published in 2003. *The Ringplatz*, where the city hall was located, was the central place of the city, its *Times Square*, and the *Herrengasse*, with its elegant fashion shops, its *Fifth Avenue*. My two uncles from my mother's side Mendel and Moritz Haber used to go to café *l'Europe* on *Herrengasse*. They were businessmen exporting wood among others to Holland, for the construction of dams. Bukowina produced and exported all over Europe a lot of wood, in particular beech. The name Bukowina comes probably from the Slavic word *buk* or the German word *Buche*, which both mean beech. In German Bukowina is also called *Buchenland*.

But Czernowitz was first of all a city of culture. According to some historians, at the beginning of the 20th century there were in Czernowitz more bookshops than bakeries. No wonder Celan called Bukowina the land populated by people and books. In Czernowitz there was a big theater hall, where regular performances of theater, opera, and classical music by local and renowned visiting companies and soloists took place. In 1847 Franz Liszt performed there, and in 1867 the famous singer Adelina Patti did too. The crown prince of Romania specially came from Bucharest to attend her concert. More recently, the renowned opera star Viorica Ursuleac who was born in Czernowitz and who would become Richard Strauss' preferred soprano, creating many of his compositions, performed in Czernowitz in the period 1923–1924. In the 1930s the famous German actress, Ida Ehre, acted there.

As far as I remember, in the cafés one would mostly hear German. At home we spoke German. My parents spoke also a little Ruthenian; they used it when they addressed the maid (until 1940 there was always a maid in our house, although my parents were not really wealthy; thus we never had our own house or flat) or when they wanted to say something that was not for my ears; at that time I did not understand Ruthenian at all. I had to learn Ukrainian and Russian in 1940, when the Northern Bukowina (and Bessarabia, which had also belonged between 1918 and 1940 to Romania), were occupied by the Soviet Union. That I had to learn simultaneously two Slavic languages, Ukrainian and Russian was due to the fact that Bukowina was integrated into the Ukrainian Republic of the Soviet Union, but Russian was the official language of the USSR. These two languages differ only slightly. That made it for me quite difficult — I was ten years old — to assimilate these languages, the more so because I did not have the opportunity to practice them. The local population continued to speak the languages they were used to. I refreshed in my mind the little Russian I had learned from 1940–1941 during my university studies in Bucharest, since in Post-War

Romania, like in all Soviet satellite countries, Russian became an obligatory foreign language. While I never reached a level in Russian sufficient to read poetry for example, I managed eventually to read physics textbooks and papers and later, in the 1960s, I gained my existence by translating these into Romanian.

Part of the Jewish population in Bukowina also spoke Yiddish. Since Yiddish is essentially a German dialect (Mittelhochdeutsch), except that it is mostly written with Hebrew characters, I understood roughly this language, too, in particular after I attended the Yiddish school in 1940–1941. Under the Soviet rule German or Romanian schools did not exist anymore. However I never really succeeded in speaking Yiddish. When somebody spoke Yiddish to me, I used to answer in German. Besides my one-year Yiddish school there are a few more circumstances when I heard Yiddish which left an impression in my memory. In chronological order they are: before the war an employee of my father used to read the Yiddish newspaper *Heint* (in Yiddish heint means today), which appeared in Warsaw. *Heint* was considered a very serious newspaper, something like the Jewish *The Times*. One day I visited my father in his wine cellar where I found him listening to a quotation from this newspaper, in which a very pertinent analysis of the political situation was made, with particular emphasis on the imminent danger Hitler-Germany represented for peace in Europe.

Another contact with Yiddish that comes to my mind is that of the Yiddish Theater. In 1945 when Czernowitz was again part of the Soviet Union, the famous Yiddish theater of Moscow led by Mikhoels gave a few representations there. I was much impressed, not only by the exceptional quality of the artists, but presumably also because that was the first time after the war that I went to a theater (like many other prominent Soviet Jews Mikhoels was killed by Stalin during the anti-Jewish campaign at the end of the 1940s).

In the ghetto I had the first contact with English and I fell immediately in love with this language, probably because it represented for me the language of freedom. A former Oxford student gave me a few lessons and after that I continued on my own, with the help of the Eckersley manuals. After 1945, when we were again allowed to own a radio set, I became a regular listener of the BBC broadcasts (this habit has continued until now, except that now I watch the BBC on TV) and that essentially completed my English "studies." I started to practice it in the 1950s, writing my physics papers (with the exception of my first two papers, which appeared in Romanian, all my physics papers and books were written in English) and then in talks and conversations, at first with foreign colleagues who were visiting Bucharest or during my first visit to Dubna in 1958, and later, after 1969, when I moved to the free world. In 1970–1972, in the USA, I gave my student courses in English and even after 1974, in Marburg, I often lectured in English, because we frequently had foreign visitors, collaborators and graduate students. Most recently I even ventured to write fiction in English and it happens quite frequently that I express myself more easily in English than in German, my mother tongue. In the first years after I moved to Paris I used English with my French colleagues, since my physics French was not yet adequate. Actually, most physics keywords used in French are anyway English, which makes the use of English often unavoidable.

The language problem was particularly serious for people with certain professions and my father belonged to that category: he had studied law at the German University in Czernowitz and Business at the Vienna Academy of Commerce, and because of the change in 1918 of the official language from German to Romanian, he could not exercise in Czernowitz his law profession. This was probably one of the reasons why he spent, after 1918, several years in Prague, where he kept a bookshop. I realize now that, given his passion for books, that has probably been for him the ideal profession. At the end of the 1920s he returned to Czernowitz, married and opened a wholesale wine cellar. He had two or three employees and this gave him enough time for reading. His interests covered most fields of intellectual activity. Dr. Max Weiner was among others a recognized expert on old languages likes Hebrew, Greek, and Latin. I remember that we had, among others, frequent visitors from the

Theology Department of the University of Czernowitz, who came to consult my father on certain unclear points in the translation of the Old Testament from Hebrew into Greek and Latin. Later he learned also Romanian and eventually was even able to translate Romanian poetry into German. My mother, Pepi, called Pupzia by her family and friends, although she had lived under Romanian rule for twenty-five years, did not speak Romanian at all.

My Family

My father's family lived in Storojinets, a small city about half an hour by car from Czernowitz. It was a big family composed of my grandparents, four sisters of my father with their families and four brothers; only one of these brothers was married and had a family. The fifth brother was living in Toulouse. He had studied there in the 1920s electrical engineering and had then settled down in France. I never met him; he died in the fifties, before I came to the West. In the late thirties I met, however, his wife Alma and their two year-old son Guy, when they visited the family in Storojinets. During the war this branch of the Weiner family survived, hidden in Southern France. I would meet again my aunt and my cousin 35 years later, when I went to Paris for the first time, during my London stay. In 1941 my 75 year-old grandmother (my grandfather had died before the war) and my father's four brothers were deported by the Soviets to Siberia. With the exception of the grandmother, who died of exhaustion on her way to Siberia, they somehow survived. I saw this part of my family only many years later. Actually, I met my cousin Rosi again only in 1989, when she, together with her family, managed to come out of the USSR. She was at that time a grandmother already. (When we had parted in 1941, she was 13 and I was 11.) In 1942 the four sisters of my father and their families were deported to Transnistria, this time by the Romanians, who during 1940–1944 were the allies of Germany. Three of them lost there their husbands and one lost there her six year-old only child.

My memories of my mother are much more reduced, because I lost her at the age of 13; she died in 1942 at the age of 43 because of inadequate medical treatment in the ghetto. I remember her as a beautiful woman who was very much loved by my father and by her (and my father's) family. She had one sister and two brothers. The sister lived in New York since World War I. I never met her or her family. She died in the late 1940s, soon after we had re-established mail contact. The younger brother, Moritz Haber, lived in Czernowitz and the elder, Mendel had moved, during the twenties to Czechoslovakia, where he married, and then he moved to Poland. At the end of the thirties he returned with his family to Czernowitz. My uncle Moritz and his wife Rosa committed suicide in 1942, when they were about to be deported to Transnistria. From a neighbor of theirs we learned that the police, who came in the early morning to fetch them, gave them fifteen minutes to prepare themselves. They went to the bathroom where they took both potassium cyanide. My mother knew that they had obtained this poison, some weeks before, from a befriended pharmacist. They had no children. I still remember my mother's desperate look when she returned from their home, where to the police had called her, to take care of the burial arrangements, so that their flat could be immediately at the disposal of some German or Romanian official. They lived on the same street — Wassilko Street — as Paul Celan's family. Paul Celan's parents were deported on the same day and perished in Transnistria. A plaque in the house where Celan's family lived reminds us about this most famous son of Czernowitz.

Menziu, his wife and their daughter survived the ghetto, but their son, in trying to escape from a labor camp in Romania, died in 1943, at the age of 19, when the boat, which was supposed to bring him to Palestine, was scuttled.

While my grand parents were religious, my parents were not. Nevertheless, presumably to please our grandparents, important Jewish holidays were kept in my parents' house. Even the religiousness of my grandparents has to be qualified. When at the beginning of the First World War my father was mobilized into the Austrian army, his father accompanied him to the railway station. Recruits were supposed to bring with them food that would resist the heat for several days (it was summer). On the way, my grandfather, who was aware that my father didn't care about religious prescriptions, stopped at a grocery's and bought for my father some smoked ham. For a practicing Jew who was not allowed to eat pork, that was a sin. My father, who had been very much impressed by this gesture, told me this story to illustrate how parents' love overcomes religious prescriptions. Unfortunately, this does not always happen.

Czernowitz had a big and beautiful temple and very many synagogues; the Nazis destroyed most of them in 1941, including the temple. However, before the war my father used to take me more often to a Catholic Church — the Polish Church — which was in our neighborhood, to listen to the organ music, which he, and later on I too, enjoyed very much. (He was also a great lover of classical and opera music, in particular that of Wagner, and my mother told me that he had chosen my first name as a tribute to this composer). The liberal attitude of my father towards religion was also reflected in his frequent contacts with Christian theologians. He remarried in 1945 and lived in Focşani, in Romania, where he worked as an accountant, a boring and badly paid job. After Vienna, Prague, and Czernowitz, the small city of Focşani was quite a change for him. Among many other things, he missed very much his books, which he had to leave in Czernowitz, and the atmosphere of the Austrian cafés, which had survived in Czernowitz also during the Romanian period of 1918–1940. My father died in 1962 in Bucharest at the age of 72.

After leaving in 1945 I visited Czernowitz once more, in 1958, with the occasion of my trip to Dubna, to meet some members of my father's family who had returned from the deportation in Siberia. I spent only two days there, but that was enough to realize that this was not the city I had known. Except for some buildings, which had survived the war, it was for me a strange Soviet city, but which triggered in me terrible memories. In the last years, due primarily to the growing interest in Celan's poetry, Czernowitz has become for Westerners a place of cultural pilgrimage; regular trips are organized, among others, from Berlin and it appears that the city has profited from these developments. Although this cannot revive what is gone forever — the city of people and books — this process is to be welcomed as a tribute to the heritage of European civilization.

My Schools

Between 1936 and 1940 I attended in Czernowitz the elementary Meisler School. This was a private school where all students and teachers spoke German, although the official teaching language, as demanded by the regime, anxious to romanize the Bukowina, was Romanian. Besides German and Romanian, from the second year on one taught also French. That I have an excellent remembrance of that school is presumably due to the outstanding level of its teachers (I remember in particular Herr Löwy and Frau Singer) as well as to the fact that the four school years I spent there were the only ones in my whole school life when, despite what happened in the world, I had no other worry but just to go to school and to learn.

In the year 1940–1941 under the Soviet regime, all children of school age in the "liberated" territories of North Bukowina and Bessarabia had to repeat the school year because, from one day to another, the teaching language had changed. Instead of Romanian one had the choice between three teaching languages: Russian, Ukrainian or Yiddish. In all schools, however, the study of Russian and Ukrainian was obligatory. Although the majority of the population still spoke German, despite the fact that according to the Molotov-Ribbentrop treaty the non-Jewish Germans had left for the third Reich, German was not taught anymore. The same applied to Romanian despite the fact that a large part of the rural population was Romanian. What concerns Romanian this arbitrary political decision had to do with the creation in the former territory of

Bessarabia of a new Soviet Republic, the Moldavian Soviet Republic, the official language of which was Moldavian. Moldavian was nothing else, except by name, than Romanian, but the rulers made a point to consider it a different language so that they could justify the creation of the new republic in occupied Bessarabia. To make this artificial distinction more believable, they decided to change the alphabet of the language from Latin to Cyrillic.

While the annexation of Bessarabia by the Soviet Union is easier to understand, because before 1918 Bessarabia was part of Tsarist Russia, for Bukowina this does not apply; this region had never belonged to Russia. According to the same imperialistic logic of the Communist regime, children in Bukowina had to study in school Russian *and* Ukrainian as main languages. To make this sudden transition easier for me, my parents sent me to the Yiddish school. Although I understood many of its words because of their German origin, I could neither speak, write, nor read Yiddish, although I knew the Hebrew alphabet used in Yiddish. (As a concession towards my grandparents, I had had before 1940, private Hebrew lessons). Still, at home, we, like the majority of the urban population, continued to speak German.

There are two things which I remember from that school year: I came in contact with the Yiddish literature and in particular with the writings of Shalom Aleichem, which I liked very much. However my Yiddish was not good enough and in order to really understand them I ended up reading them in German translation. The humor of this writer roused in me the interest for good jokes. It is perhaps at the root of the fame of Jewish jokes, which, like Shalom Aleichem's writings, very often leave you with a taste of bitterness and sadness.

The second thing which I remember of that school year, was that the teachers told us that God does not exist and that religionis an invention of the exploiting class. Although my parents were not religious, I hadn't heard before such radical statements, neither at home nor at the Meisler School, and the first thing I did was to ask my father whether this was true. My father gave me then my first lesson of philosophy. According to him, the statement "God doesn't exist" was quite primitive and essentially meaningless, as long as the concept "God" was not defined. For my father the most satisfactory definition of God was that due to Spinoza, which can be summarized by the equation: *God = Nature*. In this sense God does exist because who would deny the existence of nature? To support his point of view my father quoted some famous names like Goethe and Einstein who were adepts of this pantheistic philosophy. These names were familiar even to me. At the age of seven my father had taken me to my first Opera performance, Faust, by Gounod. To understand the story I was given to read Goethe's play, which my father knew by heart. My mother rightly objected that it was too early for me to understand this deep philosophical work.[b] In fact, I was not so much impressed by the philosophical meaning of the deal between Faust and Mephisto than by the magic tricks of Mephisto, by the ballet of the opera and by the beautiful hall of the Czernowitz theater, a vestige of the Austrian past, which I visited for the first time. Goethe's name came up, of course, also in the German class of the Meisler School, when we read some of his poems. Whether Einstein's name was also mentioned in school I do not remember. It was, however, mentioned in politics discussions among adults at home, as an example of the absurdity of the Nazi regime, which had forced his greatest scientist to leave.

Philosophy was not the only subject my father introduced me to. As a matter of fact, almost my entire education in the period 1941–1945 was homemade and due to my father. I also benefited from my father's marvellous library of several thousands books, mostly in German, which miraculously escaped the war events until 1945, when we left Czernowitz for good. The only book of his library, which my father took with him and which I still have, after so many "changes of residence," is a very old — the printing date is 1609 — edition of the Bible in four different Latin translations, one from Hebrew, one from Greek, one from Aramaic and one from

---

[b]Much later I would also understand Goethe's statement "He who has art and science also has a religion, but those who do not have them better have religion."

literary Latin in everyday Latin. My father had used this bible in his comparative translation studies. For me this book has of course much more than bibliographical value and I hope that our daughter Diana Free will continue to keep it. Among the other books of my father's library I remember of, besides the complete works of classic German writers and many volumes of English and French literature in German translation, are his chess books (he was a passionate chess player) and the complete edition of *Die Fackel*. *Die Fackel* was a famous Austrian literary journal, which appeared in Vienna between 1911 and 1936. Its founder, editor, and, from 1911, sole author was Karl Kraus, one of the greatest Austrian writers, critics and authorities of German language. My father who had spent several years in Vienna had known him personally and was a great admirer of his. I have been reminded recently about this unique personality of Viennese culture reading the autobiography of Erwin Chargaff, who had also been strongly impressed by Kraus. Last but not least I also found in my father's library a book by Moczkowski on Einstein's theory of relativity. It was a popularization book with very few formulae and served its purpose very well; it made me curious and avid to learn more about this fascinating subject. It opened for me a new horizon, that of the ever expanding frontier of human knowledge. I was first of all surprised — shocked is the proper word — by effects like contraction of length, dilatation of time, twin paradox, which sounded like science fiction. Among these strange effects there was one in particular, which impressed me most: the equivalence between mass and energy. And quite soon, I would have a concrete opportunity to guess what it really meant.

In 1943 the Jewish community of Czernowitz attempted to organize a private school with Jewish teachers. My father had taught me some Latin, German and Math, so that I was accepted into the third grade of lyceum,[c] which, however, I never finished, because the Romanian authorities closed the school after only two months. In 1944, after the reoccupation of the Bukowina by the Soviets, I

[c]The Romanian secondary school "lyceum" was based on a eight year system, which ended by a final Examination, the "baccalaureate."

did not attend school either, because children of my age (I was 14), as well as women (men were mobilized to the army), were taken to work in the coalmines of Donbas. For a few months I kept hidden and did not go out on the street, because there were raids and women and children were sent directly from the street to the labor camps. For this reason, when we moved in 1945 to Romania, our first stop was Suceava, a small town in Southern Bukowina, where I passed, in two months, some equivalence examinations, which entitled me to enter the sixth grade of the lyceum. In Suceava I also had an experience which would play a decisive role in my future life.

In 1941 the Soviets, before leaving Czernowitz, had requisitioned and then burnt all the radio sets to insulate the population from any contact with the outside world. (After the war this Soviet disinformation policy would continue by jamming Western broadcasts). During the following ghetto period under Nazi-Romanian rule Jews a fortiori were not allowed to possess or even listen to radio broadcasts. The first time we had again this opportunity was after war in 1945, when we moved to Romania. During the two months spent in Suceava we started to listen again to the BBC news broadcasts. In the small pension where we lived there was a radio set in the sitting room, where guests used to gather. One evening, it was on the famous August 6, 1945, the broadcast was interrupted and the speaker announced that the US Air force had dropped an atomic bomb on Hiroshima. There were not many people in the room who understood what this meant, but everybody was shocked. My father explained to me that this had to do with the relation between mass and energy of Einstein. This impressed me very much, and since I could not follow the details of this explanation, it remained recorded in my mind as one of the (many) things I would like to understand in my future studies.

2

Politics — Premonition of War

In 1918 when Bukowina was taken over by Romania, two thirds of its population were non-Romanians and did not speak Romanian. If only for this reason, the change of regime was not welcomed by the majority of the population. For Jews in particular, who had no citizen rights in Romania, the takeover was a serious political and social setback, not to mention the cultural aspects of this change. Even after 1923, when Jews were granted citizenship, anti-Semitism was in Romania more than in most other European countries quite a common phenomenon. Already at the age of six or seven I came to realize that belonging to the "chosen people" was not necessarily a privilege. Indeed, the events, which happened in the late thirties, were so disturbing for Jews that even a six-year-old child like myself could not avoid taking notice.

I remember that in 1936 or 1937 I had been visiting my father's family in Storojinets. Because I had in Storojinets so many cousins almost of my age and with whom I could play, I used to come in the 1930s to that place with great pleasure. One evening I found in the living room of my grandparents all my uncles listening to the radio. I usually did not pay attention to emissions that interested adults, but judging from the faces of my uncles something exceptional seemed to take place. It was a speech by Adolf Hitler transmitted live on the radio. I found it unusual firstly because of the hysterical shouting of the speaker and the noisy applauses of his audience in Germany, who were scanning for minutes "Sieg Heil" and "Ein Reich, ein Führer." But after I heard some of the sentences of Hitler's speech I understood why the adults looked so much concerned. Hitler said that all the evils of this world were due to the Jews and that the German people cannot be saved but by eliminating the Jews. After a few minutes my uncles had enough of it and switched to the German broadcast of the BBC. At this moment I also understood why my parents usually listened to the German news broadcasts of the BBC or radio Beromünster (Switzerland) and not to that of stations from Germany. This event made a deep impression on myself and was probably the motive why London had been the city of my dreams for almost forty years. I also remember street manifestations, which took place in Czernowitz in the late thirties, when mobs marched through the streets shouting "Goga, Cuza"; these were the names of two extremist politicians who were competing in nationalism and anti-Semitism with the Iron Guard Legion — a straightforward fascist organization — and who had been appointed by the King to form a government in a last attempt to oppose the takeover of Romania by Nazi Germany. One of their first decisions was to strip Jews of their citizenship.

Two years later I realized concretely what war meant when I saw in September 1939, soon after the invasion of Poland by Nazi Germany, thousands of Polish refugees, in cars or in carts drawn by horses, streaming through Czernowitz on their way to the South. Moreover, at that time a Jewish refugee from Poland lived in our flat for some time. Because of the war events she had been separated from her husband and hoped to meet him in Czernowitz, but this never materialized. She learned later that the Germans had deported him to Majdanek. Her stay in Czernowitz was presumably not legal, because I remember that my parents hid her in a "secret" room. In front of the door to that room, a cupboard was placed and the tiny window was walled up, so that at a first glance one did not notice the existence of the room. What happened next to that woman, I don't know.

These events were a premonition of what would happen in less than one year. In 1940 the Soviet Union, with the agreement of Nazi Germany, occupied Northern Bukowina and Bessarabia. This was termed by the Soviet propaganda "liberation" from capitalism. The occupation of North Bukowina and Bessarabia in 1940 by the Soviet Union was one of the arbitrary acts of the Molotov-Ribbentrop agreement, through which the two "socialist" regimes, USSR and Nazi-Germany — the Nazi regime called itself national-socialistic — under the pretext of fighting "capitalist imperialism," extended their own empires and reshaped the political structure of Central Europe. Other consequences of this dictate were the division of Poland between Germany and the Soviet Union, the occupation by the USSR of the three Baltic States and the annexation by Hungary of most of Transylvania, which, like Bukowina, had been until 1918 part of the Austrian-Hungarian Empire and then attributed by the Versailles Treaty to Romania.

For most of the population of Northern Bukowina the occupation by the Soviet Union was a step backwards. Besides the change of political regime and of the official language with all its cultural and administrative consequences, it implied also a substantial deterioration of the standard of living, which in the USSR was, at that time, much below that of Romania. Because of the inefficiency and absurdities of its state controlled economy, the Soviet Union endured an acute shortage of consumer goods and lodging. Although only ten, I realized certain changes in our daily life, due to this state of affairs. At first certain goods, like citric fruits and bananas, which used to be imported, disappeared completely. Moreover, many of the Soviet newcomers didn't even understand what we were missing; they had, apparently, never seen these things. When they were asked why there were no lemons, they answered: "This is a temporary transportation problem, our country has many factories (sic) producing lemons." It took some months until the new regime managed to "socialize" completely the economy. During this transition time certain private enterprises like small restaurants still continued to function. When the newly arrived Soviets entered such a restaurant they were amazed to find here a menu, from which one could choose. They were even more surprised that they could order as many courses as they liked. While in capitalist countries the poorer people did not have enough money, in the Soviet Union money was, apparently, not a problem; the problem was that there was nothing to buy with that money. Therefore in the restaurants, orders by the emaciated Soviets of omelettes made of ten eggs became standard. They coined even a name for this omelette, "desiatka," from the Russian *desiat,* meaning ten. Two steaks or two complete menus per person were also customary. After a month or two things "normalized" and the chronic shortage of everything reached also our place. However soon we would learn that in the Soviet Union, too, there was a privileged class, the natchialniks (the Russian word for higher officials and privileged party members), known later as the *nomenklatura,* who had access to special shops, where one could buy almost everything.

Housing, too, was in the USSR a severe problem. Moreover, the standard of Soviet flats or houses was apparently so low that a bathroom was considered a capitalistic privilege which a Soviet citizen could do without, and some of the newcomers had to learn how to use water closets.

These aspects of life became, however, secondary, due to other, more dramatic events. In May–June 1941 the Soviets deported all citizen who had passports with the index "39" to Siberia. In the Soviet Union "passports" were not travel documents, as in most other countries, but just identity documents. Travel was restricted even within the Soviet Union. Travel to a foreign country was a luxury a normal citizen could not dream of. Passports with index "39" had been issued to people considered by the Soviet authorities as "bourgeois," that is socially "unhealthy" and politically unreliable. My parents also had "39" passports, but we managed to escape deportation. I remember these circumstances quite well, probably because for the first time in my life (I was only eleven) I felt like I was on the brink of death.

In an attempt to "improve" his social status, my father became in 1940 a blue-collar worker. He worked manually in a factory producing quilts. One day a factory colleague of my father warned him that the NKVD (the precursor of the KGB) had made inquiries about us. According to Soviet customs this meant that our arrest and deportation was most probably imminent. My father suggested that we try to hide until this "wave" passed. My mother, however, was reluctant, in part because I was sick (I had mumps). Anyway, in the evening we left our home (arrests happened usually during the night) and took the tramway to the river Pruth. My parents intended to take refuge in a suburb, across the river, where they knew somebody. Somehow that did not work out and we had to return to the city. In going again over the bridge, my mother, in her despair, proposed to my father that we all three jump into the river and "bring the story to an end." My father managed to convince her that they couldn't do that, if not for other reasons, but because they had no right to decide about my life. They brought me to a father's cousin, where I remained a few days. I had high fever and had to stay in bed. (I presume that my parents were afraid to bring me to my mother's brothers, who also lived in Czernowitz and who did not have "39" passports, because the NKVD would look for us there as well). Where my parents hid I do not know anymore, but eventually all three of us escaped deportation. During these days tens of thousands of people from Bukowina were deported, among them half of my father's family from Storojinets. A few weeks later the Soviets withdrew. With the occupation of the Bukowina by the German and Romanian troops a new chapter of my "adventurous" life started.

Making Ends Meet

The period 1940–1941 and a fortiori the following war years were for our family quite a difficult time also from the material point of view. In 1940 the wine cellar of my father was "nationalized" and we lived from the wage my father gained as an unqualified factory worker and from selling our household. The first thing we sold was the piano. I had taken four years of piano lessons with a private teacher. In 1940, I went to the Soviet music school in Czernowitz and made my home exercises at a neighbor's who had a piano. In 1941, with the beginning of the war, this activity, of course, stopped. What we lived from in the period 1941–1945 I don't know anymore; I vaguely remember that my father started to work as an accountant. His income was very modest and there were periods when we did not know, literally, how to make ends meet.

In 1944 or 1945, after liberation from the ghetto, my mother's sister from New York managed to contact us and sent us some parcels. This was a great help; from selling the content of such a parcel, we could live a month. One day I got a letter from her with a one-dollar banknote enclosed; it had escaped censorship. In the USSR possessing hard currency was illegal, but we decided nevertheless to keep it as a reserve. When we crossed the border to Romania, I hid it in one of my shoes. As a matter of fact, it was the only reserve we had and when we arrived in Suceava where I passed my high school equivalence examinations, we exchanged it there on the black market and bought food for a few days.

3

War — The Ghetto

On July 5th, 1941 Romanian troops accompanied by German SS units entered Czernowitz. Immediately after that the Jewish temple was burnt down and thousand of Jews in Czernowitz and the rest of the Bukowina were executed, among them the representatives of the Jewish population and the Chief Rabbi of the Bukowina, Dr. Mark. Then the Jewish population of Czernowitz was evacuated into a ghetto. About 50,000 people, who had lived all over the city, were concentrated in a few streets. I remember that in the room of about 15-square meters where we lived for weeks, there were no fewer than 25 people, who could lie down only in shifts.

After that began the deportations to Transnistria, a region in the Ukraine, East of the river Dniestr. Three quarters of the Jewish population of Czernowitz were deported. And Czernowitz was a fortunate exception, because in the rest of the Bukowina the entire Jewish population shared this fate. Of those deported, the majority perished in the camps. At a certain moment the deportation from Czernowitz stopped. The authorities had apparently realized that the activity of the town was about to collapse. Certain people like us were allowed to stay. This was, of course, only a temporary measure and had the evolution of the war been different the rest of the Jewish population would have ended up in the camps, as well. As a matter of fact, our stay in Czernowitz during the Nazi occupation was everything but peaceful. Like in the Middle Age, Jews had to wear a yellow star, of a prescribed dimension. Apart from those who went to work and who had special permits, Jews were permitted to go out to do shopping for two hours only, from 10 a.m. to noon. Every person, children included, were issued identity cards with a yellow star, which they had to bear with them on all occasions. I happen to still have this document and it reminds me of another circumstance, when I felt that I was again near to death. It was in 1942 or 1943, my father worked and my mother was bed-ridden. One morning I went to buy bread and to lend from a friend some books for reading. It happened that on that morning the authorities organized a raid and together with many other people I was taken to the police headquarters, where I was kept until midnight, without being told why. The rumor among the arrested was that the Germans had asked the Romanian authorities to speed up the deportations and picking up people from the street was simpler than fetching them from their homes. Actually, many of those arrested were deported next day to Transnistria. I was saved from deportation by a neighbor of ours, Dr. Hilsenrath, who was at that time the president of the Jewish community and who had in this function some official contact with the police. Another detail of quite different nature I remember is that when I was arrested I had with me two breads, which I had just bought, as well as some books I had borrowed from a friend. Thus, during my arrest I (and the people around me) had what to eat — I am not sure whether we were given water to drink — and what to read. If the circumstances would not have been those mentioned, one could hardly find a better illustration of the Roman dictum *panem et circenses*.

Writing now, for the first time, after almost seven decades, about these two episodes of my childhood in Czernowitz during which I narrowly escaped deportation and most probably death, they appear to me in hindsight almost as natural events. Of course, this sentiment is deceiving: hadn't these events left traces that cannot be erased, I wouldn't have remembered them after so many years. Although when I left Czernowitz I was too young to draw parallels or analogies between the two dictatorship regimes I had lived through, I was conscientious that the occupation for the second time in 1944 of Czernowitz by the Soviets, despite all its negative consequences, distinguished itself fundamentally from that in 1940: It saved my life and was an act of liberation, indeed. Had it happened a few weeks later, it would have come too late for us. The preparations for the deportation of the rest of the Jewish population from Czernowitz were quite advanced. Actually, such a last minute deportation took place in Hungary and cost the life of hundreds of thousand Jews. Not only was I too young in 1945 to understand what really had happened — actually, like many others, even today I have difficulties in this respect — but at the moment I left Czernowitz for good I did not suspect that the sentiment of freedom I experienced at last would be of short duration, that I was about to enter a new and much longer chapter of life, again under a communist dictatorship, that of communist Romania.

Section II

Post-War Romania; The Isomeric Shift; Persona Non Grata (1945–1969)

4

High School and University

Focşani 1945–1949

Although the reoccupation of the Bukovina in 1944 by the Red Army had saved our lives, the first thing we thought about once we were again under Soviet rule, was how to escape this rule. The opportunity for this to happen appeared for the first time in 1945, when former Romanian citizens, who lived in the Northern Bukowina, were allowed to move to Romania. Our first stop was in Suceava, a small town in the Southern Bukowina, where I passed some high school examinations and then we moved to Focşani, another Romanian town more to the South, where three sisters of my father, who had survived the camps in Transnistria, had momentarily settled.

From 1945 to 1948 I went to *Liceul Unirea* in Focşani. Focşani was at that time a city of about 60,000 habitants. It had played in 1859 an important role in the unification of the two Romanian lands, Moldova and Walachia, and *Liceul Unirea* (unirea means in Romanian unification) traces its name back to this event, in honor of which it was founded. Economically Focşani is the center of one of the main wine producing regions of Romania and that was one of the reasons why my father decided to settle there; he hoped he might be able to renew his contacts there and find a job.

Being one of the oldest in the country, *Liceul Unirea* was a school of some renown. Among the teachers I had there I remember professor Arbore, the teacher of Romanian language and Romanian literature. At that time my knowledge of Romanian was quite limited and I had no practice of it. In Czernowitz, even under Romanian rule, I almost never used it, the less so under the Soviet regime. Professor Arbore realized that Romanian was still a foreign language for me. As a matter of fact, among the 30 or so students in my class, I was the only one in this situation. To my surprise after one year only, he quoted me as an example of a good student in Romanian. I think that one of the reasons for this rapid change was my forced experience with languages. The main reason, however, was probably Professor Arbore himself. Although or perhaps just because he was considered as the strictest of the teachers the school had, he was an excellent teacher. Last but not least, from the very beginning a kind of affinity had been established between the two of us, due to the fact that he, too, spoke several languages and in particular also German. He was extremely knowledgable in universal literature and philosophy, and there was no lecture when he did not quote, from memory, in the original language, one of his preferred writers or philosophers, be it Shakespeare, Goethe or Kant. I later found out that before 1945 he had in fact been a university professor, but for political reasons had not been allowed to remain at the university after the war. I realize now how lucky I was to have had him as my teacher. Actually, after the sixth grade I had to choose between the scientific and literary sections of the lyceum and that meant choosing also between Professor Arbore, who was teaching in the literary section and some other teacher of Romanian literature, who was teaching in the science section. Although at that time I had already decided to study sciences, I nevertheless opted for the literary section, just to continue to have Arbore as my teacher. The curriculum in the two sections differed mainly by the fact that in the literary section one continued to learn Latin, while in the science section one had mathematics. I learnt therefore, on my own, in parallel, math, and at the end of the year, I passed an equivalence examination and entered the eighth (and last) grade of the science section. My passing the baccalaureate subsequently as the best student of my class, with the highest grades ever obtained in that school at this examination, Romanian language and literature included, is in part due to this eminent teacher.

Looking back now it seems to me that besides my father, who taught me so many things, in particular at the time when I did not have the possibility to go to school, it was Professor Arbore who influenced me most in my adolescent years and I will always remember him with gratitude. I was very saddened to learn that because of his "unhealthy social origin" (his family had owned large estates), two or three years after I had left the lyceum Professor Arbore had been arrested and deported to the labor camp of the Danube-Black Sea channel[a] (he was at that time in his fifties, already, and suffering from a heart condition), where he soon perished.

Once we arrived in Romania we intended to leave as soon as possible; for us it was a foreign country and we were not welcome there. Romania being at that time a satellite of the Soviet Union, people who had left the Soviet "paradize" were viewed as traitors of socialism. Nevertheless getting out of Romania was next to impossible: actually it took me 24 years until I succeeded to cross the Iron Curtain and even then only illegally. More than any other socialist country with the exception of the Soviet Union, Romania turned out to be a prison for its citizens.[b]

In a first, unsuccessful attempt to leave Romania I joined in 1945 the Zionist youth movement. For a displaced person like myself, the Zionist idea of a Jewish state (which at that time did not yet exist) offered at least a perspective — all the more so given that the long tradition of anti-Semitism, which had found its

[a]The Danube-Black Sea Channel was one of the first demented construction projects of communist Romania. Later on Ceauşescu would bring this megalomania to a new level by further insane projects at the expense of the needy population.

[b]Yugoslavia for example had in this respect from the beginning a very liberal attitude; Hungary, Poland and Czechoslovakia followed up to a point this example.

climax in World War, II,[c] continued, albeit in other forms, under the communist regime. Strangely enough, even after the Ceauşescu regime, when most of the Jews had left for Israel, anti-Semitism continues to make headlines in Romanian politics.

After World War II Palestine was mostly a desert and the youth Zionist organizations prepared their members to become pioneers and to live in collective agricultural settlements (Kibbutz). Although I, personally, did not feel vocation for this kind of life, the urge to leave prevailed and in 1948, after finishing the lyceum I went to an intensive preparation course in Tg. Neamţ, a city in the Northern part of Romania, where, one learned, among other things, Hebrew. The State of Israel had just been proclaimed and the Soviet Union and its satellites were among the first countries which recognized it. This made us hope that Romania might allow Jews to emigrate. This hope turned out very soon to be illusive. In fact, a few days after my arrival in Tg. Neamţ, in the autumn of 1948, the Central Committee of the Romanian Communist party, in one of its "historical" resolutions, decided that Zionism was a tool of Western imperialism, which had to be combated by the progressive and peace loving forces. Consequently, Zionist organizations were forbidden and I had to return to Focşani. (My Zionist adventure had an after play to be mentioned below.)

Since it was too late to register for the entrance examinations — only students who passed these examinations were admitted to university — I had to wait until the next academic year. During this year spent in Focşani I read a lot, trying to find out what university studies suited me best. I was initially interested in philosophy, but I realized quite soon that in a communist country philosophy meant just Marxism-Leninism. Besides that it seemed to me that there was no way of doing philosophy without a solid scientific base. The

[c]During the years 1941–1943, the vast majority of Jews from Bessarabia and the Bukowina were killed, either on the spot, when the Romanian troops re-entered in 1941, or after deportation to Transnistria. On top of that Romanian troops, allied with the Germans, killed tens of thousands of Ukrainian Jews during the war.

nearest thing to do was to study sciences of nature, in particular physics. In high school I liked math very much, which I studied mostly on my own, because the math teachers I had were not very good. (That applies for physics as well, but in this case I would grant my teachers the benefit of doubt: the physics labs of the school were quite primitive). I decided to present myself in 1949 to the admission examinations of the Department of Mathematics and Physics at the University of Bucharest, which I passed easily.

Bucharest

Moving to Bucharest was for me a memorable event. Besides being a university town, Bucharest was a center of culture which had a lot to offer to a young person like myself, who had been deprived during the war, not only of almost everything a human being needs for daily existence, but also of the cultural life and the contact with the world, which I longed for. Czernowitz had been before the war a city of remarkable culture and in some sense also an international city and I missed its atmosphere. The four years spent immediately after the war in Focşani — a small, uninteresting, provincial city — were very far from that.

Unfortunately, in my first two student years I could not afford to participate in many of the cultural events, which took place in Bucharest, although tickets were usually quite cheap. I had barely enough money to buy my daily lunch and dinner in the student cafeteria, and to pay the modest rent of the room I shared with my friend Levy Rahmani. As a matter of fact, we had even to share the bed — the room was too small for two beds. It was an almost completely dark room, its only window leading into a corridor. The truth is that, as far as I remember, I didn't care very much: My studies not only absorbed me completely, but after the forced interruption during the war and later in 1948, they offered me sufficient intellectual satisfaction to compensate for any material want. Levy, at present a renowned psychologist in Israel, felt the same. Our sharing of rooms continued also later, when we could afford to have two separate rooms: We kept the first room, which had the advantage that it was better heated in winter and which we baptized therefore the "winter palace," and shared a second room as the "summer palace." Eventually when we found a more convenient substitute for the winter palace, Levy moved into this room, while I continued to keep the "summer palace," which was of about the same size as the "winter palace," with the difference that it had a true window. It is this room that became later my private "institute," Labirint[d] 28 (cf. below).

Despite the fact that at the end of our common Bucharest period, which for Levy ended at the beginning of the 1960s, we did not share the rooms anymore, we continued to meet once or twice a week. This relationship was suddenly interrupted in the mid-fifties because of a dramatic event, which ended for Levy almost in tragedy. Although it happened more than fifty years ago, that event shocked me so much that I remember it as if it had happened yesterday.

Our rooms were in the same district, but Levy's room was slightly more central than mine. Therefore on my way to or from the city, I passed quite often by Levy's place and, if not in a hurry, I used to call him to the window (his room was on the first floor) by whistling the main tune from Beethoven's overture Egmont. It happened that one week he did not show up at the window for several consecutive days. Since this occurred also on an early Sunday morning, a time he always used to be at home, I became worrisome and, because he had no telephone, I went up the stairs and rang at his apartment. In that apartment lived, besides him, another family I knew superficially from my previous visits to Levy. I rang the bell two short times, the agreed signal used by Levy's visitors. To my surprise the door was opened by his neighbor who, after checking

[d]Labirint was the name of the street. It owed its name to its winding shape.

that there was nobody in the stairway to see us, called me in and told me that the Securitate had arrested Levy three days, actually three nights, before; as usual they had shown up at 4 a.m. The neighbor was still under shock and this shock took hold of me as well. A few days later I learned that during that night several former leaders of the Zionist youth organizations had been arrested. Levy had been a prominent member of Hanoar Hazioni and although I myself, two years younger than him, had not been as prominent, I could well imagine that I might be on the next list of arrests. The following two weeks I did not sleep at home but at my uncle's, Muniu, the most junior brother of my father, who had managed to escape in 1946 in an adventurous way from his Siberian deportation, and who had moved in the 1950s to Bucharest. (Muniu left at the end of the 1950s for Israel and would play, in 1969 an essential role in my escape from Romania.) This hiding of mine was of course a very naïve reaction: in hindsight both Muniu and I realized that had the Securitate intended to arrest me too, they would have found me easily. Levy and his comrades were kept in prison two years before they were bought out by the Israeli government and permitted to leave for Israel.

University

The university was for me what physicists call a *change of phase* and I enjoyed very much the transition from high school to university. What I had mostly disliked in high school was the obligation to study things that did not interest me, like zoology and botany, probably because they were presented as an enumeration of dry facts. Another subject I found boring was the early development of Romanian literature, which started by translations of the bible. I remember in this context "Psaltirea" by Coresi, an important oeuvre for the history of Romanian language and literature, but extremely insipid for a teenager in the middle of the 20th century. History and geography of Romania did not interest me either. This lack of interest for local things was in sharp contrast with my interest for facts connected with foreign countries and the more remote the country, the bigger my interest for it. That tendency started already in elementary school when my schoolteacher, Herr Löwy, used to tease me for not knowing the most important cities or rivers of Romania, but knowing all important cities and rivers of North and South America. As a matter of fact, the only subjects which I was really interested in were mathematics and philosophy. I disliked physics in school, among other things because the experiments performed in the lyceum were very rudimentary and also because the teacher presented them very dryly. For the same reason I disliked chemistry (taught by the same teacher), but also because you had to learn many formulae by heart. This learning by heart was one of the most unpleasant duties of my school years. That is why I also never considered studying medicine, although the subject itself attracted me.

My interest in philosophy arose due to my father's lesson on pantheism, the meaning of which I interpreted as nature governing itself by its own laws and that to understand these laws you have to study sciences of nature. I soon realized that it was no accident that in the past philosophers had also studied science. However, the time when one individual could study all scientific disciplines had passed long since; I had to make a choice. Still, I was pondering between mathematics and physics. While it seemed to me that physics was more related to what I understood at that time by philosophy, my interest in mathematics was more of esthetical nature. I considered solving a mathematics problem a kind of amusement, comparable with that of reading an interesting book or contemplating a work of art, in other words mathematics was more fun. Actually, the difference between mathematics and physics is much deeper: physics is a science of nature while mathematics is rather a science of the mind.

Since at the University of Bucharest the departments of Mathematics and Physics were at that time not yet separated, in the first year the courses were common. This suited me because it meant that I could postpone for another year my choice between mathematics and physics. I found the introductory physics courses in the first university year quite uninteresting, the more so the laboratory experiments, (the equipment was, as in high school, completely outdated) and I did not meet any interesting physics professor. The mathematics courses were of much higher quality, in particular the course of analysis by Miron Nicolescu, which I (and most of my colleagues) considered absolutely brilliant. This was in part due to the fact that in Romania mathematics had a much older tradition than physics, presumably because physics is an experimental science, which needed closer contact with industry. And in pre-war Romania industry had been quite underdeveloped. Still, at the end of the first university year I decided to continue my studies in physics, because I had realized that it interested me more. In hindsight I think that to this decision contributed, besides the philosophical arguments discussed above, the two biographical events I mentioned before: a) my contact with Einstein's theory of relativity during the auto-didactical period in the ghetto; b) the atomic bomb, an unwanted consequence of this theory.

Among the physics courses, which I appreciated most in the following three university years, I remember those by Richard Woinarovski, Remus Răduleţ, Josef Ausländer, and Şerban Ţiţeica. Woinarovski lectured on Analytical Mechanics and from him I got for the first time the notion of physical approximation, which I consider essential for a physicist. Răduleţ gave the fundamental course on electrodynamics, which included the theory of special relativity, and Ausländer lectured on chemical physics, a dry and difficult subject, which neither the lecturer (he was a nuclear physicist), nor the students liked very much, but which Ausländer was able to present in a very logical and condensed way. Josef Ausländer (his friends called him "Jusziu") was also from Czernowitz and had made his physics studies in Berlin and Zürich; because of his Jewish origin and of the political events he had to interrupt his scientific work and returned at the end of the 1930s to Czernowitz. In 1940 he got, at the request of Einstein, who had written on his behalf to the US ambassador in Bucharest, an entry visa to the USA, to join the Princeton Institute. Unfortunately this invitation came too late; only a few days after he got the visa the Soviets occupied Bukovina and the curtain (it was not yet called the "Iron Curtain," but it was not less opaque) was drawn. After the war Jusziu came, like me, to Bucharest. With the help of Hermann Hesse, whose wife was a cousin of Jusziu, he managed to leave Romania at the end of the 1950s. After some time spent at CERN at the invitation of Vicky Weisskopf, he joined the Kernforschunbgszentrum and the University of Karlsruhe. He died prematurely in the 1980s.

Last but not least, of my Bucharest physics professors I have to mention Ţiţeica, who was considered by everybody *the* professor of Theoretical Physics. *I* attended his lectures on Statistical Physics and Quantum Mechanics. Ţiţeica's courses were so pedagogical that the students often had the impression that he proved things, which in reality could not be proven and which were, rather, conjectures. This is true, in particular for his "derivation" of the postulates of Quantum Mechanics in the spirit of Heisenberg's matrix approach. I have been amused by a remark by Steve Weinberg in his book "Dreams of a Final Theory" where he comments upon what seemed to him this mysterious derivation. Had Steve attended Ţiţeica's course, he might, perhaps, have thought differently.

Ţiţeica had been in Leipzig one of the first (if not the first) PhD students of Heisenberg; he formulated in his PhD thesis the first quantum theory of the Hall effect. After returning to Romania in the 1930s he continued to work on Quantum Electrodynamics. I understand that in the 1940s he had solved the renormalization problem but, because of the war events, did not publish his results. After the war he realized that his results had been obtained, independently, by another student of Heisenberg, Tomonaga, who, however, had published them and who was awarded for these results the Nobel Prize. Subsequently, possibly out of disappointment, Ţiţeica practically ceased to do research (at least he did not publish on his own anything worth mentioning except his courses on statistical physics and quantum mechanics), but had quite a few PhD students who became theoretical physicists. Some of them left Romania in the sixties and seventies and settled in the USA (Hardy Mayer), Western Europe (Mihai Gavrila, Viorel Sergiescu, myself) and Israel (Chaim Iussim, who changed his name in Yatom).

In 1950–1951 the Romanian Academy was reorganized along the lines of the Soviet Academy of Sciences. New institutes were created, among them the Institute of Physics. In 1951, at the end of my second university year, several students of the Physics Department, me included, were appointed to the Institute of Physics as "preparators." This constituted essentially a fellowship, which gave us the possibility to finish our studies without doing extra jobs to earn our living. In 1953 I got my Physics Diploma (something intermediate between a master degree and a PhD) with a thesis on scattering of photons on electrons, more precisely the derivation of the Klein-Nishina formula, under the supervision of Ţiţeica. That gave me the opportunity to get acquainted, on my own, with relativistic quantum mechanics and the gamma matrix algorithm. (The regular courses did not go beyond non-relativistic quantum mechanics). After that I was appointed researcher at the Institute of Physics. In 1958, I got from the University of Bucharest my PhD (at that time it was called, as in the Russian system, "candidate in science."[e]

My connection with the University was renewed ten years later when I was appointed "senior physicist" at a newly created Computing Center of the University. How this happened will be explained later on.

[e]This name originates in pre-revolutionary Tsarist Russia. The university system of pre-war Romania had been, however, different, actually quite similar to the French system. In this system the Doctor title was equivalent to the British or American PhD. Satellite countries like Romania showed initially their unconditional submission to the Soviet Union by adopting not only "socialist" habits, but even specific Russian ones. This was done in the name of the fight against "imperialism." Later on, under the Ceauşescu regime, the "socialist brotherhood" was replaced by a nationalistic policy, reflected also in a return to pre-war Romanian traditions. Among other things the Doctor title was reintroduced and assimilated with the candidate title.

Theoretical Physics

In the second year of my physics studies at the university it became clear to me that I was more interested in theoretical than in experimental physics. This was in part due to my initial interest in philosophy and to the deplorable state of the physics laboratories of the university. At the newly created Bucharest Institute of Physics of the Romanian Academy, there was, from the beginning, a Theoretical Physics division. However, because of personal animosities between some professors, the very existence of this division was also more of "theoretical" nature, at least during the period 1953–1958. Initially most of the junior theorists, myself included, belonged to the group headed by Andrei Popovici, whom we had not met before at the university; he was a newly appointed professor and a high-ranking party member. The rumor was that his party merits were his only credentials, but as a student I could not judge: He was working on unified field theories of Kaluza-Klein type, at that time a topic inaccessible to me and apparently to most of his professor colleagues, with whom he was at odds. Whether this conflict was due to his political position or to his scientific views I don't know. His archenemy was Horia Hulubei, the director of the Institute of Physics of the Academy, an atomic physicist of some fame due to his previous work in France.[4] Hulubei's career in post-war Romania was in part bolstered by the support he had from some leading French physicists, among them Joliot-Curie and Yvette Cauchois. Joliot-Curie, besides being an eminent scientist was also a leading figure in the French communist party. Due to this fact his influence in the scientific life of post-war communist Romania, with its traditional ties with France, was quite appreciable. Eventually Hulubei succeeded in getting rid of Popovici who lost his position at the Physics Institute of the Academy. A few years later Popovici committed suicide. It is unclear to me whether this was due to his loss of political influence or the lack of understanding his work had met among his colleagues. If a scientific giant like Boltzmann had been driven to this gesture, it wouldn't have been surprising that a Popovici would have had the same fate. It is a sad irony that Kaluza-Klein theories, Popovici's "fad," have become in the last decades a highly respectable subject. The members of Popovici's theory group became associates of various experimental divisions and I landed — without being asked — at the laboratory of optical spectroscopy led by Ion Agărbiceanu, professor of physics at the Bucharest Institute of Technology ("Polytechnic Institute"), where I was supposed to continue to do theoretical physics on my own.

Being aware that I felt somehow lost and to encourage me, Agărbiceanu told me the story of his PhD studies. Like the majority of prominent Romanian physicists of that time, Agărbiceanu had made his physics studies in France and got his PhD at the famous Laboratory of Aimé Cotton. Upon his arrival at that laboratory he was given a key to an empty room and was told that he was expected to assemble his experimental apparatus himself, by taking or lending various bits and pieces from the laboratory shop or from colleagues. And most importantly, he was "of course" supposed to choose himself the subject of his research. Recently I have found in the memoirs of Abragam, who had done his physics studies in France during that same period of 1930s, similar observations about the French PhD system. In hindsight I think that this system has many positive aspects, although not everybody enjoyed that "freedom."

Ion Agărbiceanu was one of the very few Romanian physicists who, after their PhD in France and before returning to Romania, had also worked in the UK. I always associated his elegant appearance — he used to wear for instance in summer immaculate white linen suits — with this British chapter of his life. He had enjoyed very much his stay in England and regretted his return to Romania. He told me many interesting and funny stories about this country and that increased even more my interest in it. One of the stories I still remember refers to his landlady, who was convinced that in her house there was a ghost who used to make his appearances during the night.

I felt great sympathy for Agărbiceanu and I think that this feeling was reciprocal. I considered him a perfect gentleman in every sense of the word. To this contributed his democratic convictions, (his father had been a famous *progressive* writer, which was rather exceptional in pre-war Romania), our common admiration of the British for their steadfastness during the war and, last but not least, the fact that, despite the difference of age and position, he treated me as a colleague. We met several times a week and used to discuss physics and non-physics issues.

It turned out that I too followed Agărbiceanu's example of choosing for myself the subject of my PhD thesis, although initially I was not so happy about this privilege. A few years later, while I was busy with my thesis, the Institute of Physics, located in Măgurele, a suburb of Bucharest, was divided into two institutes: an Institute of Atomic Physics, under the directorship of Hulubei, which remained in Măgurele, and an Institute of Physics headed by Bădărău, and located in the city. Both Hulubei and Bădărău had been professors of mine at the University and both were members of the Academy. As it happened before, I (and some of my colleagues) landed again, without being asked, in another place, this time at Bădărău's institute. The theory division of this institute had no leader and I continued to be on my own, a state of affairs I was already used to.

5

The Isomeric Shift on Spectral Lines

The Discovery of the Isomeric Shift

Unlike most physics effects — the Mössbauer effect and the isotopic shift effect to be discussed below included — the isomeric shift was at first predicted and then found in experiment. In trying now to reconstitute the circumstances under which this prediction came about, three factors come to my mind:

I. My interest in finite size effects. Already in high school I had been preoccupied by questions like: how elementary are "elementary particles"? Is there an end to the quest for the smaller and the still smaller?

II. My interest in line widths of spectral lines stemming from my study of the energy-time indeterminacy relation and my affiliation with the optical spectroscopy laboratory of Agărbiceanu.

III. The emergence, a few years before, of the nuclear shell model.

Finite Size Effects in Subatomic Physics

The "planetary" model of the atom in which electrons are moving around the nucleus is an illustration of the fact that the atom is not, really, what its Greek name a-tom, i.e. indivisible, suggests, but rather an object of finite size. This is, of course, just the beginning of the story, because, except for the nucleus of the hydrogen atom, the nuclei of all other atoms are compound systems, made of protons and neutrons, and have, thus, finite sizes, too. (The nucleus of the hydrogen atom, being a proton, was considered initially as "elementary" and thus without finite size, but the electron scattering experiments of Hofstadter, not to mention earlier developments, made us abandon this simplistic picture in that case as well). The finite size effects of the nucleus have been seen for the first time in atomic spectroscopy through the isotopic shifts of spectral lines. For the better understanding of what would follow I would like to review a few definitions:

Nuclear isotopes are nuclei, which differ by their number of neutrons, the number of protons, i.e. the electric charge of the nucleus, being the same. Since the chemical properties of the elements are determined by the number of electrons, which (for neutral atoms) is equal to the electric charge of the nucleus, isotopes of a given element have the same chemical properties and occupy the same place in the periodic system of elements. Here from the name isotope (in Greek *isos* means the *same* and *topos* means *place*).

The *isotopic shift of spectral lines* is the shift in the atomic spectra, which occurs when we substitute one isotope by another.

Nuclear isomers are nuclei in which the number of both protons and neutrons is the same (*meros* in Greek means *part*), but which differ by their quantum states, characterized by energy and angular momentum. With respect to the periodic system nuclear isomers are also isotopes, because they, too, occupy the same place in this system.

In analogy to the isotopic shift, the *nuclear isomeric shift* is the shift in atomic spectra, which occurs when one replaces one nuclear isomeric state by another.

While the notion of isotope is used only in (nuclear) physics, the notion of isomer is used also in other fields. Thus isomers in meteorology are lines on a map connecting points having the same percentage of precipitation. In chemistry isomers are substances that have identical molecular formulas, but different molecular structures or configurations, and hence different chemical properties.

Natural Line Width and the Limits of Optical Spectroscopy

The optical spectroscopy laboratory of Agărbiceanu at the Institute of Physics had a Fabry-Pérot interferometer, a quite powerful instrument, providing high resolution in the visible domain of atomic spectra. With this instrument Agărbiceanu's group studied among other things the influence of pressure of the emitting system on the position and width of atomic spectra. I contributed, as a theorist, to one of the experimental publications of this group[5]; this constitutes the only experimental publication in my career as a physicist. However the short period I spent in Agărbiceanu's laboratory probably had a lasting influence on my future scientific career, since it oriented me towards phenomenology, a direction I would follow for the next fifty years.

The topic of optical spectroscopy made me wonder about the ultimate limit of the experimental methods used: this limit is represented by the natural line width, that is the (inverse of the) lifetime of the excited state, the decay of which leads to the emission of the spectroscopic line. To my surprise I found that this ultimate limit was related to one of the obscure topics of non-relativistic quantum mechanics, the energy-time indeterminacy relation. This relation, although treated in most textbooks as an analog of the momentum-coordinate indeterminacy relation, has a different meaning, because time in non-relativistic quantum mechanics is not a physical variable, which characterizes a given system and thus changes from system to system, as space coordinates or other physics variables do, but has a meaning independent of the system, it just "runs" independent of its surroundings, "il coule et nous passons" in the words of Lamartine.

Mathematically this peculiar property of time is reflected in the fact that unlike other physics variables time is not a hermitian operator. That time is not an "observable" in non-relativistic quantum mechanics manifests itself also in the irreversibility of certain thermodynamical phenomena; the measurement process itself, which is fundamental in the interpretation of non-relativistic quantum mechanics, is also an irreversible process.

As a consequence of the different interpretations of time there are at least two different interpretations of this relation, one in which time refers to the *duration* of the measurement and another where time indicates *the moment* when the measurement takes place. The first interpretation leads directly to the natural line width because this is the lower bound for the duration of the measurement. Since I was not satisfied with the way this relation is discussed in the literature, I wrote a review paper, which appeared in 1955[6] where I attempted to clarify the state of the art at that moment. As a matter of fact, I believe that this state of the art has not changed since then, as our understanding of the privileged role time plays in quantum mechanics has not much changed in the last fifty years. It is conceivable that this mystery will find its solution with the understanding of extra dimensions in string theory, because in a certain sense time, too, is an extra dimension in quantum mechanics. This was not only my first independent publication but also my first encounter with the limitations of an analogy in physics. However that did not deter me from attempts of analogies of my own, which are the subject of this book.

The Fabry-Pérot interferometer could also be used for the investigation of other effects that interested me, because they were related to the fundamental question of finite size effects mentioned above: It permits the measurement of very small differences in atomic levels. Such small differences are due to the hyperfine structure[7] (the name itself is suggestive) of atomic spectra and to the isotopic difference of atomic nuclei. This last effect leads to the isotopic shift of spectral lines: The atomic spectra of two isotopes of the same element are shifted one with respect to the other, despite the fact that the electric charges of the two isotopes are identical. Indeed, although isotopes differ one from another only by the number of neutrons, their electromagnetic interaction with the electrons of the respective atoms are different, because the masses and volumes of two isotopes are different.[a] The difference in volume is clearly a finite size effect.

Let us consider now the case of two isomers, which have the same number of protons *and* the same number of neutrons, but which still differ by their quantum state, i.e. by energy, spin, magnetic, and electric moments. Spin or angular momentum is the physical quantity, which characterizes the rotation properties of a quantum system. If the system has electric or magnetic properties, external electric or magnetic fields will influence its rotation. The strength and direction of this influence is measured by the electric and magnetic moments. The fact that, because of the difference in the magnetic (dipole) and electric (quadrupole) moments, the hyperfine structure of two isomers must be different was obvious. But at a first look one might think that for isomers there would be no shift of spectral lines, analog to the isotopic shift, because the number of neutrons, which determines the difference between two isotopes and thus the isotopic shift, is the same for isomers. *One day, however, I asked myself whether this was really so: did not, because of the difference of the two quantum states, the charge distributions of two isomers differ?* What I did not realize at that moment (but what I became convinced

[a]The first theory of the volume isotopic shift was proposed in 1931 by Pauli and Peierls. When I met Peierls in Oxford in 1973, he was interested to learn about the relation between isotopic and isomeric shifts, a topic he had not followed for a long time.

a few months later) was that with that question the isomeric shift had been found.

The answer to the above question was not trivial, given the fact that the excitation of the nucleus and the phenomenon of isomerism are effects due to the strong interaction, which are also responsible for the stability of nuclei, while the charge distribution is by definition an electromagnetic property. Moreover, it was conceivable that even if the nuclear charge distributions were different, the difference would be too small to be accessible to experiment. Furthermore, like any researcher plagued by a new idea — I went through these doubts also on later occasions, when working on virgin soil — I asked myself why had not somebody thought of the isomeric shift before; nuclear isomers were known for at least twenty years.

As a matter of fact, the question "why not earlier" has been asked in the case of the Mössbauer effect, as well. Referring to this effect Gonser[8] writes: "Retrospectively, one might be surprised that this effect was not discovered earlier. The application of elementary quantum mechanics both to nuclear and solid state physics has been understood for many years. It seems that the failure to realize the basic principle of this effect earlier was the lack of interaction between the nuclear physics and solid state physics communities." For the Mössbauer effect the question "why not earlier" is even more justified, because this effect could have been observed *experimentally* earlier, while for the isomeric shift the technique had yet to be developed.

The only theoretical physicist in Bucharest whom I considered competent enough to discuss these questions with was Şerban Ţiţeica and he was sceptical. He thought that even if the charge distributions of two isomers were different, the difference would be too small to be observable. At that moment I thought that this was, possibly, also the reason why nobody had addressed this issue in the literature. I would soon find out that there was another explanation for this delay.

Anyway, I did not give up; I wanted to convince myself whether this effect existed and if so, whether it was measurable. (Here, too, one might see a similarity with the Mössbauer effect. After

Mössbauer had published his paper, some researchers in the US could not believe his results and redid the experiment just to verify it). To answer my question about the difference of charge distributions, I had to make myself familiar with theoretical nuclear physics, a subject that was not taught at that time at the university. I had to understand what were the mechanisms that determined the very existence of different isotopes and isomers, their stability and/or their lifetime.

A colleague of mine, Dragoş Bogdan, also a theoretical physicist and with whom I shared the office, was working at that time on beta decay. This effect involves transitions between isotopes and isomers and from Dragoş I found out that the most successful nuclear model, which dealt with the questions that interested me, was the nuclear shell model proposed in 1949 by M. Mayer in the USA and J. Jensen in Germany. This model was so recent that it was not yet treated in textbooks and I had to go to the original literature.

Atomic Versus Nuclear Shells, the Pauli Exclusion Principle and the Nuclear Shell Model

I mentioned above the analogy between the isotopic and isomeric shift. In the prediction and estimate of the isomeric shift another analogy played an important role, namely the analogy between atomic and nuclear shells. This analogy was for me (and, as I would find out soon, not only for me) a surprise. Indeed, atoms are bound states, consisting of one nucleus and one or more electrons, the binding being due to the electromagnetic force. Nuclei, on the other hand, are systems consisting of nucleons, bound primarily by the nuclear force these two forces are quite different. The electromagnetic force is long ranged, while the nuclear force is short ranged. Furthermore, the strengths of these forces are quite different: the electromagnetic force is two orders of magnitude weaker than the nuclear force. Last but not least, in atoms there exists a natural reference point, the nucleus, whose mass is much bigger than the mass of the electrons. No such reference point exists in nuclei.

So, at a first look these systems are totally dissimilar. Still, they behave similarly in the sense that both have a shell structure, which leads to quite pronounced periodicities in their properties. These periodicities manifest themselves in both cases through the fact that closed shells have higher binding energies and thus are more stable than unclosed ones. The periodic system of Mendeleyev, which explains the chemical properties of elements, is nothing else but a reflection of the shell structure of atoms. Among nuclei the periodicities, which explain, among other things, the stability and lifetimes are reflected in the existence of certain special numbers of nucleons called *magic*.

The explanation of this fundamental similarity between atoms and nuclei lies in the fact that in both cases we deal with fermionic systems, governed by the Pauli principle, which excludes the possibility that two identical fermions should be in the same quantum state. The effect of this exclusion principle is so strong, that it overshadows all differences between electronic and nucleonic systems. When I first learned about the difference between fermions and bosons, which is also reflected in the different statistics properties of these two species of particles, I was perplexed. After more than fifty years I continue to be perplexed. The reason for my bewilderment is that I have not been able to find any *intuitive* explanation for this strange fact, which is a postulate of quantum mechanics. Although the relation between spin and statistics can now be derived from quantum field theory, its intuitive explanation is still elusive (confer, however Chapter 31).

The Isomeric Shift and the Shell Structure of Nuclei

The analogy between atomic and nuclear shells was a surprise for me. However, what surprised me even more was the fact that it leads to a remarkable simplification of the calculation of the isomeric shift. While at a first look the problem I was faced with was a complicated many body problem involving all nucleons, it turned out that there was a class of isomers for which, in a first approximation, it was sufficient to consider one single nucleon, called the "optical" nucleon, to get an estimate of the difference between the charge distributions of the two isomer states; the rest of the nucleons were *filtered out*.[9] In a second publication[10] on this effect I described this state of affairs by stating: "The comparison of charge distributions of two different states may supply new information on the excitation process. This comparison is automatically realized in nature by a subtraction process in the nuclear isomeric shift on spectral lines." It was this last feature, which had intrigued me most. After all, even if — to use Einstein's words — "nature is not malicious," it seldom is that forthcoming. The class of nuclei I was referring to were isomers in odd proton-even neutron nuclei, near closed shells. In-115, for which the effect was calculated in Refs. 9 and 10 is such an example. The result of the calculation was that the nuclear isomeric shift should be measurable by optical spectroscopy: although rather small, it turned out to be two orders of magnitude bigger than a typical natural line width.

After I had convinced myself that the effect "existed," I went to the library and searched for the literature to see which experimental groups had done atomic spectroscopy with isomer nuclei (that the isomeric shift had not been mentioned in the experimental or theoretical literature, I had checked previously). I found one single experimental group in this category, that of Francis Bitter of MIT (Bitter became later the first director of the National Magnetic Laboratory of the USA, where the strongest magnet of the world was built) whose team had developed a new optical and magnetic scanning technique, which could be used for this purpose. Bitter and co-workers had published in 1953 and 1954 two papers along these lines, where they reported about their experiments on the hyperfine structure of Hg-197, and where they had measured the spin of the excited isomer state.

I wrote immediately to Bitter, told him about my estimates and asked him whether he wouldn't be able to look for the isomeric shift, preferably in odd proton-even neutron isomers (like In-115) where the theoretical predictions were more reliable. I also submitted for publication a *letter to the editor* of Nuovo Cimento, at that time the most respected European journal of high-energy and nuclear physics.

To my pleasant surprise both Bitter and the editor of Nuovo Cimento answered by return mail. Bitter told me that he had been extremely interested in these results and that he and his collaborators hoped to be able to look for this new effect. However, for technical reasons, they would start with Hg-197, the element they had worked on already before.

I was very happy about this prompt and encouraging answer, although I realized that Hg-197 was not the best element to look for the isomeric shift. I remember that the secretary of Bădărău, the director of the Physics Institute, handed over Bitter's letter to me, and it happened that he entered the secretariat room when I opened the letter. I had been, apparently, smiling, because he asked me why I was so amused. I showed him the letter and explained to him that I had the impression that the Americans were looking for the key, not exactly where it had been lost, but rather where the light of the street lamp was. Bădărău, apparently, did not share my doubts: As recounted to me later by Moisil, a mathematician who was also a member of the Academy and who would play a role in my escape from Romania, Bădărău went straight to the Academy to boast with the achievements of one of the members of his institute. It turned out that Bădărău was right after all. The shift measured three years later in Hg was of the same order of magnitude as that predicted for In and, as shown soon after that by Shirley,[11] this could be explained by the theory developed in my papers (cf. below).

As to the editor of Nuovo Cimento, he accepted immediately the paper for publication and it appeared in December 1956 under the title "Nuclear Isomeric Shift on Spectral Lines."[9] Some years later I learnt from Alexandru Sanielevici, one of my former professors at the Bucharest University, that he had met in Italy Radicatti, the prominent Italian theorist who told him that he had been the referee of my paper, that he had liked it and had recommended its publication without delay. About twenty years later I met Radicatti at CERN and he still remembered my paper.

With the prediction of the isomeric shift my PhD became a simple and pleasant formality. Since it was a theoretical investigation I asked Ţiţeica whether he would agree to be my thesis advisor and accept the isomeric shift as a subject for my thesis, which he did. I wrote up the thesis and presented it officially to the faculty of Mathematics and Physics in 1958. After the unanimous vote of the faculty, one of my former professors, Miron Nicolescu came to me and congratulated me with the words: "you have grown."

Realizing that my idea was taken seriously, I published a more detailed paper about the effect in 1959 in *The Physical Review* under the title "Charge Distribution of Excited Isomeric Nuclei and Atomic Spectra,"[10] in which, among other things, I calculated the isomeric shift for Au-197 (this calculation has been extended for the Mössbauer isomeric shift by Shirley in reference 11) and studied the influence of the form of the nuclear potential on the charge distribution.

In the formalism outlined in Refs. 9 and 10 the isomeric shift is a product of two factors: (1) The wave function of the electron inside the nucleus, and (2) the difference of charge distributions in the two nuclear states. The electron wave function had been calculated in the 1930s, since it appears also in the isotopic shift. Therefore the calculation of the isomeric shift amounted essentially to the calculation of the difference of charge distributions. As described above, for nuclei near closed shells, this difference could be easily estimated from "first principles."

Very soon after that Melissinos and Davis from the Bitter group reported in Ref. 12 their results. They found an isomeric shift in Hg-197 of the same order of magnitude (2×10^{-2} cm^{-1}) as that predicted in Ref. 9 for In-115. This was an important result, since it constituted the first experimental confirmation of the existence of the effect.

From a theorist's point of view Hg-197 was not an ideal element to study the isomeric shift, because, unlike In-115 or Au-197, the odd nucleon is a neutron.[13] The interaction between an electron and a *free* neutron is much weaker than the conventional Coulomb electron-proton interaction and therefore, at a first look, one might expect for odd-neutron nuclei a much smaller isomeric shift than for odd-proton nuclei. However, as pointed out already in Ref. 9, what was at stake in a nuclide like Hg was the interaction between electrons and *bound* neutrons and this demanded special considerations. As a matter of fact, five years later Shirley[11] suggested that the effective electron-bound neutron interaction reduces the effective electron-bound proton interaction only by the factor Z/A. Using the Au-197 result, this lead to an isomeric shift in Hg-197 quite close to that found in experiment.

The key role which the shell model played in the evaluation of the isomeric shift explains also why this effect was predicted only in the fifties and not twenty years before, when isomers were discovered: only after the shell model had been formulated, an estimate from first principles of the charge distributions became possible.

The approach to the nuclear isomeric shift suggested in Refs. 9 and 10 was later applied also by other workers in the field, for instance by Lardinois.[14] The Mössbauer isomeric shift is at present formulated under the same form as that used in Refs. 9 and 10.

The analogy between the isotopic and isomeric shift is based on the fact that both are due to a difference in the electromagnetic interaction energy between the electrons and the nucleus of an atom. The isotopic shift had been known decades before the isomeric shift and it played an important role in the early days of nuclear physics. It must be stressed, however, that while the isomeric shift is sensitive to the internal structure of the nucleus, the isotopic shift is (in a good approximation) not. Therefore the nuclear physics information, which can be obtained from the investigation of the isomeric shift, is superior to that which can be obtained from isotopic shift studies. The measurements through the isomeric shift of the difference of nuclear radii of the excited and ground state constitute one of the most sensitive tests of nuclear models. Moreover, combined with the Mössbauer effect, the isomeric shift constitutes at present a unique tool in many other fields, besides physics.

Some Confusion of Terminology

In 1956 when the notion of isomeric shift was coined and the effect was predicted I was aware of the possible confusion between the isomeric shift in spectral lines due to *chemical* isomers, and that due to *nuclear* isomers. That is why I proposed for the latter the name *nuclear isomeric shift*,[9] to be distinguished from a possible *chemical isomeric shift.* At that time, of course, the Mössbauer effect was not known and the isomeric shift could be studied only by atomic spectroscopy methods. In the subsequent physics literature slight variations of terminology are used. Thus in the paper by Melissinos and Davis[12] where the first experimental evidence for the nuclear isomeric shift was reported (by using methods of optical, i.e. atomic spectroscopy), the notion of *isomeric isotope shift* is used and, what is even more disturbing, in the papers where the Mössbauer effect was used to measure the isomeric shift, one finds also the name *chemical* shift.[b] Last but not least, in Ref. 15 the name *atomic isomer shift* is used for what was defined in Refs. 9 and 10 as the nuclear isomer shift and the name *nuclear isomer shift* is used for the shift seen in nuclear (Mössbauer) spectroscopy, when one compares the same nuclear transition in two atomic systems, which have different electronic wave functions. Clearly, this last definition refers to the *experimental method* by which the isomeric shift is observed, rather to its *cause*.

One reason for this chaotic terminology is the fact that immediately after the Mössbauer effect had been discovered, there was a kind of gold-fever atmosphere in the physics community and

[b]This is due to the fact that in Mössbauer spectroscopy the nuclear isomer partners are in different chemical configurations.

there was no time for such trivial things like searching the literature. Another reason was the lack of interaction between people working in atomic, nuclear, and solid state physics. The following quotation from the talk by Shirley at the Second Mössbauer Conference reflects this state of affairs[11]: "I would like to remind you that the isomer shift was first observed not by a Mössbauer experiment but in an optical spectroscopy experiment by a group at MIT about two years ago.[12] They observed an isomeric shift in the optical transitions in two isomers of Hg-197 and got a very reasonable value, which is quite close to what one might predict on a very simple model. Apparently optical spectroscopy can be used to study isomeric shifts for quite a few other nuclei. This is very useful because it is complementary to the Mössbauer effect. A simple theory was worked out by Weiner to explain this kind of effect, and our calculation is actually quite similar to his theory, except that it applies to the Mössbauer resonance instead of the optical transition."

Even this clarification leaves room for further clarification, since in Ref. 11 the reference (with date and journal) to the papers by Weiner is missing and because of that the unadvertized reader may not understand that the "simple theory worked out by Weiner" preceded the experimental observation and thus predicted the effect.

The Mössbauer Effect

Two years after the publication of my first paper on the isomeric shift on spectral lines, the fundamental discovery by Mössbauer that the emission of the gamma ray by an excited isomer is under certain conditions, recoilless, appeared in the literature,[16] and in 1960, the first measurement of the isomeric shift with the help of the Mössbauer effect was reported.[17] It is this combination of the two phenomena, known at present under the name *Mössbauer Isomeric*

Shift, which triggered an avalanche of applications (cf. below) and which made this effect so famous.

The nuclear isomeric shift seen in atomic spectroscopy depends on the electron wave function near the nucleus and on the difference of the nuclear radii of the isomeric and ground state. Now suppose that we consider a given nuclear isomer in two different physical or chemical environments (say different physical phases or different chemical combinations), which, because of the experimental arrangement, we shall call source and absorber, respectively. Then we will have different electron wave functions in the source and the absorber, and thus, on top of the nuclear isomeric shift discussed above and which is due to the difference of the nuclear states, there will be a shift between the two environments. This combined shift is called at present the "Mössbauer isomeric shift" and it is described mathematically by the same formula as the nuclear isomeric shift, with the exception that instead of one electron wave function of the source, one deals with the *difference* between the electron wave function of the source and the electronwave function of the absorber. By measuring this shift one obtains important and extremely precise information, both about the nuclear isomer states and about the physical, chemical or biological environment of the atoms, represented by the electronic wave functions.

The Mössbauer isomeric shift can be measured with the help of a resonance phenomenon by tuning continuously the energy of the source. The tuning is done by simply moving the source with respect to the absorber. This motion produces a shift called Doppler shift, which depends on the relative source — absorber velocity. Resonance is achieved when this Doppler shift compensates the isomeric shift and makes the absorption line coincide with the emission line.

In Mössbauer spectroscopy there are thus two types of Doppler effects. There is the Doppler effect, which, through the recoil of the source, accompanies any emission process. Then there is the Doppler effect induced by moving the source with respect to the absorber. To optimize the resonance effect, i.e. to enhance the precision of the measurements based on it, the Doppler shift due to the recoil has to be as small as possible. Mössbauer's discovery consists essentially in the observation that such a minimization can be achieved by lowering the temperature of the environment in which the nuclear source is embedded. This makes possible the use of the Mössbauer shift for measurements of displacements and energy levels with experimental means that are not only much simpler than those of conventional atomic and nuclear spectroscopy, but also many orders of magnitude more precise. In the Mössbauer isomeric shift the extreme sensitivity of the Mössbauer effect comes into its full bearing. The extraordinary precision of this method can be seen by pointing out that one measures relative energy changes smaller than 10^{-13}, which are converted into displacements per unit of time. The length scale, which would have to be used to achieve in measurements of the relative changes in distance to the center of the moon the same accuracy as that, which is obtained in the measurements of the relative energy changes, is just one micron. The high precision in the measurement of relative energy changes comes about through the fact that the isomeric shift is a shift in atomic levels of the order of 10^{-5} eV, which has to be compared with the energy of nuclear levels, which are of the order of 100 MeV. If one is interested in the "environmental" aspect of the problem, the nuclear part, i.e. the difference of nuclear radii, is treated simply as a given parameter, which is determined empirically. Alternatively, in the nuclear physics investigations, the electron wave functions are treated as parameters, or calculated from theory.

Under its Mössbauer variant, the isomeric shift has found important applications in domains as different as Atomic Physics, Solid State Physics, Nuclear Physics, Chemistry, Biology, Metallurgy, Mineralogy, Geology, and Lunar research. Without the Mössbauer effect the isomeric shift would never have reached its present fame. However the reciprocal is also true: if the nuclear isomeric shift had not existed, in other words if the difference of the two nuclear radii had been too small to be measured, most of the present applications of the Mössbauer effect were not possible.

According to one of the reviewers of this field, "the isomer shift might be considered the most unique feature of the Mössbauer effect."[18] This symbiosis of two different physical effects is, in itself, quite a unique phenomenon.

For reasons which are explained below, after 1959 I did not participate in the further developments of the isomeric shift, neither in optical, nor in Mössbauer spectroscopy. In particular, I did not attend any of the conferences dedicated to this subject and did not have access to the proceedings of these conferences. Furthermore, because of the adventurous circumstances under which I left Romania, my correspondence referring to this chapter of my scientific career was lost. Only in 1977–1978, during a sabbatical at Los Alamos, I had the time and opportunity to browse through the literature of optical and Mössbauer spectroscopy and to get in contact with some of the people who had been working in this field. With this occasion I found out that some of the early actors in *optical spectroscopy* knew my papers well (this was, presumably, the reason why Shirley[11] did not bother to mention the exact coordinates of my papers) and considered them as "fundamental."[19] According to Steudel[19] the Heidelberg group had discussed them intensively, but to their regret they did not have, at that time, the technical possibilities to do the optical experiments. In *Mössbauer spectroscopy*, however, the situation was different. Many of the workers in this field, including those who used to review the subject, confessed that they had not known my papers[20,21]. However, once they saw these papers (with a delay of more than 20 years), they tried to put the record straight. Here is a quotation in this sense from the introduction by Ruby[22] to the volume "Mössbauer Isomer Shifts," a volume of more than 900 pages dedicated entirely to this subject. (Just the size of this volume indicates the extension, which this topic has taken in the literature). This paragraph was introduced at the request of the editor of the volume, S. K. Shenoy, when the volume was already in print:

"It should be pointed out that the term 'nuclear isomer shift' was first used by Weiner to describe shifts in optical spectral lines from the difference of the nuclear charge distribution of isomeric nuclear states. This work calls attention for the nuclear structure information contained in the monopole interaction."

Last but not least, it should be mentioned that the nuclear isomeric shift has been observed also in muonic atoms, that is atoms in which a muon is captured by the excited nucleus and makes a transition from an atomic excited state to the atomic ground state, in a time which is short compared to the lifetime of the excited isomeric nuclear state. Like in Mössbauer spectroscopy, here, too, not many people were familiar with the history of this effect.[23]

Dubna — 1958

Very soon after getting my PhD I was invited to the Joint Institute of Nuclear Physics (JINR) in Dubna. Here I met among others Bogoliubov who was at that time the Director of the Theory Division and thus my official host, Blokhintsev, the Director General of the Institute, Puontecorvo, Smorodinski, and of course my good and old friend Hardy Mayer, who was spending a longer stay in Dubna and who had arranged my visit. Hardy was also from Czernowitz and his biography was quite similar to mine, with the exception that he had had the bad luck to be deported, during the period 1942–1944, together with his parents to Transnistria. As far as I remember I met him for the first time in Bucharest at the Mathematics and Physics Department in 1949–1950, although, as it turned out later, he had attended in 1940–1941 in Czernowitz the same Yiddish school as myself, albeit one class ahead of me. When I met him in Bucharest he was already assistant professor. Before his Dubna stay he had published a monograph on field theory, one of the first textbooks devoted to this fundamental subject.

All four Soviet physicists mentioned above, and in particular Smorodinski and Puontecorvo, showed interest in my work on the isomeric shift: Smorodinski himself had worked on the theory of isotopic shifts and Puontecorvo had done pioneering experimental work on artificial nuclear isomers. In the good tradition of his teacher Fermi, Puontecorvo was an outstanding theoretical physicist and a very talented experimental physicist as well. He was working at that time on neutrinos and had put forward the bold hypothesis of neutrino oscillations. He didn't live to see it brilliantly confirmed.

I had quite a long discussion with Puontecorvo, who asked many questions not only about my work, but also about my biography. Because of his Jewish background he was quite interested in the fate of Jews of Central Europe during the war. He asked me about what was going on in Romania and tried to compare the situation in Romania with that in his own country, Italy. All over all he made the impression of an unhappy man, an impression confirmed by Erwin Friedlander who met him a few years later. In 1958 Puontecorvo's wife was already in a psychiatric clinic, where she apparently was "treated" because of her disappointment with the Soviet regime. She had been a convinced communist and it was due mainly to her that he deflected from Britain to the Soviet Union.

After my seminar Blokhintsev offered me a long-term appointment at the JINR. This was a very attractive offer, both scientifically and financially, but to the surprise of Blokhintsev and of many of my colleagues and friends, I asked for a few months time of reflection and eventually I did not accept this offer. Two circumstances contributed to this decision of mine. The first was the general impression the Soviet Union made upon me in 1958, two years after the bloody crushing of the Hungarian Revolution. The month I spent in the USSR with the occasion of my Dubna visit confirmed my previous two years Soviet Union experience (in 1940–1941 and 1944–1945) and convinced me definitively that this was not a country I wished to live in. Although the takeover of the power by Khrushchev was followed by a political "thaw" in the interior of the country, and although 13 years had passed since the end of the war, the ferocious dictatorship with all its consequences was still present in almost every detail of daily life, not to mention the material misery in which the population lived. If I, as a foreign visitor enjoying many privileges, saw and felt it, it was easy to imagine what the day by day life of the average citizen looked like. To illustrate this fact I can mention a small detail, which, however, made a strong impression on myself.

I was spending a weekend in Moscow with Hardy and his wife Rita. We stayed at a big, modern hotel (I think it was Hotel Peking, but one could hardly distinguish it from the other recently built Moscow hotels). Common to these hotels was their bombastic, typically Stalinist style and the corresponding decorations. One couldn't avoid the impression that Stalin had the ambition to compete with his "colleague" Hitler not only in crimes, but in bad taste as well. There were analogies all over the place...

More interesting, however, was the architecture of the interior, in particular the reception hall. Actually each floor had its own reception hall. Here you had to deposit the key when leaving the room and from here you fetched it, when arriving. The reception consisted of a desk behind which there was seated an employee. From her — it was always a woman — one could also get at any time boiling water for tea. The remarkable thing about this reception desk was its position on the corridor. It was placed so that the lady could watch the doors of all rooms. Still, this supervision was not perfect. Indeed, when I returned in the evening I found in my room a few pages with some poems by Pasternak, typed on a typewriter. This famous Russian writer, who only a year before had been forced to decline the Nobel Prize awarded to him for his novel Dr. Jivago, which he had published abroad, could not publish his poems in the USSR and his supporters circulated his work by "samizdat."[c]

The Anti-Western orientation of the Soviet Union and of its satellites, imposed under Stalin, manifested itself among other things in an almost complete travel ban to the West. Scientists were among those most affected by this ban. Nuclear physics, however, was of particular importance for the Soviet regime, both for military reasons as well as for energy production. The role of the JINR was to try

[c]Samizdat is the Russian word for self-publishing.

to counterbalance this self-imposed isolation. The JINR had been conceived and built in the 1950s to provide physicists of Eastern Bloc countries with the possibility to do research in nuclear physics. It was the successor of the "Hydrotechnical Institute," which had been devoted to secret nuclear research. In the 1950s this kind of research continued in another place (only very few people knew exactly where) and the JINR had been declared an "open" institute of unclassified nuclear research, to which also foreigners had access. It was an artificial island within the Soviet Empire and a kind of showroom for the outer world. Although it had an accelerator with an energy comparable to that existing in other institutes in the West, the experimental work done here could not compete with that done in the USA or at CERN. With the exception of satellite and weapons research, this was probably true for all other areas of experimental research in the Soviet Union, reflecting the low technical level of the country.

This was in particular also true for the testing of my ideas about the isomeric shift. Indeed, my visit in Dubna confirmed what I had already suspected: the only place where hyperfine studies with radioactive nuclei could be performed was the US and in particular Bitter's laboratory. This was of course a deception. Whether this contributed to my subsequent decision not to accept the job offer in Dubna I am not sure anymore, given the fact that the level of theoretical physics research in the Soviet Union was, on the contrary, very high. Indeed, more shining "stars" than Landau, Pomeranchuk, Tamm, Bogoliubov, Zeldovitch, Sakharov to name just a few of the leading Soviet theoretical physicists, could hardly be found in any other corner of the world. I still have a vivid impression of the seminars of Landau and Pomeranchuk, which took place weekly in Moscow. Landau's seminar was devoted to all subjects of physics, while Pomeranchuk's, which followed Landau's, after a break, dealt with gravitation only. This break between the two seminars, during which the participants exchanged information and opinions impressed me most, because it reminded me of a stock market, where physics ideas were continuously bought and sold.

At this seminar I remember to have also met Zeldovitch and Lifshitz. The meeting with Lifshitz was quite amusing. Some weeks before, I had submitted (from Bucharest) a paper to the *Soviet Journal of Experimental and Theoretical Physics JETP*. The editor of this journal was Lifshitz. JETP used to publish at first in Russian, in the Soviet Union, and then the papers were translated and published in the USA by the American Physical Society. Papers had to be submitted in Russian. Since at that time I read physics papers and physics books quite frequently in Russian, (books in other languages were not available, but many Western science books were translated into Russian) I had thought that my Russian was good enough to *write* a physics paper. When I met Lifshitz I did not intend to ask him about the status of the paper, since I did not imagine that Lifshitz, as Editor in Chief, would know about it, the more so that I was aware that the subject of the paper (isomeric shift) did not interest him. But to my surprise, when Lifshitz learnt my name, he remembered immediately the paper and told me (we were conversing in English) that he had found the paper "very interesting," but there were some minor unclear points. He asked whether I couldn't send him a copy of the paper in English; he, personally, would translate it into Russian. He remembered even some of the "unclear points": I had confused the words *perturbation* with *excitation,* since in Russian these two words are quite similar (perturbation is in Russian "vosmuschienie" and excitation "vosbushdienie"). However the two words played a key role in the paper. Isomers are nuclear states, which differ by their excitation level. On the other hand to calculate the isomeric shift in the atomic spectral lines I had used perturbation theory for the electronic wave functions. My illusions about my knowledge of Russian language suffered a severe blow after this conversation. (I knew that my colloquial, not to mention my literary, Russian was very poor). Only the absolute need — I later made my living by translating from Russian to Romanian — changed this situation. As a matter of fact, the seminar I gave in Dubna was in English and in all conversations with my Russian colleagues I used exclusively English. The paper in question was soon published, at first in the Russian edition, and then (in 1959) in the American one.[24] I must confess that I was not very unhappy about this linguistic experience. After all, there were not many physics authors, not to mention at the age of 28, who could boast themselves with a paper translated by a scientist of the calibre of Lifshitz. With all due respect to the proportions, this story reminded me of the submission to *Zeitschrift für Physik* by Bose, in 1924, of his paper on statistics, written in English. Bose had sent the paper to Einstein and asked him to arrange for its publication, in case he considered it sufficiently meritorious. Einstein personally translated it into German (Zeitschrift für Physik published at that time only in German) and the paper was soon published with a note signed by Einstein about its importance. A short time after my 1958 Dubna visit I was forced to improve my Russian, not because I intended to join the Dubna Institute, but, on the contrary, because I decided to apply for emigration from Romania to the West. What an irony of life.

6

Persona Non Grata

Applying for Emigration and Its Consequences

I mentioned that one of the reasons why I decided not to join the Dubna institute was that I could not conceive to spend the rest of my life under a communist regime. Although Romania was also a communist country and thus a prison for most of its citizens, this prison did not seem to be as impenetrable and definitive as the USSR and this was the second reason why I did not accept the Dubna offer. Indeed, just before my visit to Dubna, an important event on the emigration front had taken place: *application forms* for emigration had become again available at the militia (the equivalent of police in communist Romania).

I realize that this story sounds Kafkaesque and surreal and writing it now, in the 21st century, I myself find it difficult to believe: That you cannot apply for something because the necessary forms are not available. This had been for many years (since 1951, if I remember correctly) the official and only answer people got when they expressed their wish to leave the country.

Anyway, this time tens of thousands of people (mostly Jews and Germans) used this opportunity to try to get out and I intended to do the same. That is why I did not sign the employment contract with the JINR and after my return from Dubna I also applied for emigration. Only a small percentage of applicants got the much desired *exit visa* that is the permission to leave. (The exit visa is an invention of the communist regime. For travel abroad, besides a passport and an entry visa of the country you wanted to go to, you needed also an exit visa without which you were not allowed to leave and which specified the country to which you were permitted to travel.) The majority, myself included, were denied this elementary right.

Actually, nobody ever got a formal negative answer, but you received another kind of "acknowledgment": applicants were either fired from their jobs or retrograded. This was possible because the entire economy was a state economy, under the control of the government, i.e. of the Communist Party. This situation continued for many years (in my case for almost 11 years): You were not allowed to work and you were not allowed to leave either.

Initially there was no logical explanation for this punitive action, except, perhaps, that the authorities, surprised by the big number of people who wanted to leave, decided to discourage in this way those who had not yet applied. The fact that this measure was then applied for such a long time has probably a simple explanation: the regime realized that this was a method to obtain labor at a lower cost. If the entire country was in some sense a concentration camp, where nobody, except the *nomenklatura*, was allowed to leave, another concentration camp within this concentration camp was created, in which the conditions of life were even harsher. And when, eventually, after many, many years the Romanian authorities decided to allow people to leave, they did it only because of the political and economical pressure from the West. In fact, a flourishing trade with the West, primarily Israel and later Western Germany came into being: people were exchanged for hard currency.

It is remarkable that the policy of retrogression and firing from jobs was applied by Romania a long time before it was imitated by the Soviet Union. On the other hand Romania — as far as I know it was the only European satellite of the USSR to do this — imitated the Soviet Union by organizing in the early fifties labor camps for the construction of the Danube-Black Sea canal, one of the many absurd projects of the "socialist" economy. To these camps were sent, besides criminals, also former members of "reactionary" pre-war political parties, "capitalists" and other unwanted persons. For the "political" prisoners this happened without any formal juridical process, not to mention that the entire communist juridical system was, anyway, a mockery.

A few weeks after my application for emigration I got a letter from the administrative director of the institute, in which I was notified that my position as researcher had been transformed into that of a "technician" and that my salary had been cut by more than half. Actually, the secretaries in the institute had now a higher salary than myself. I was kept in this position during the following ten years, until 1968. As to my new duties, nobody knew initially what they were, except that I was ousted from the theory division and given a desk in the draftsmen room. (I was the first, and for some time the only, scientist of the institute who had applied for emigration). Later the administrative director, who actually represented the party within the institute, wanted to use me in the bookkeeping service, which I refused. Eventually I was left alone. Besides the reduction of my salary there were also some other "changes." While the scientific staff, in particular those in the theory division, had no fixed daily schedule, I had now the same schedule as the administrative employees. Moreover, I was not allowed to go to the library during the working hours. The Physics Institute had no library of its own, but used the library of the Physics Department of the university, which was in another place. At a time when Xeroxes did not yet exist this represented for a theorist quite a punitive measure. Actually, when some years later Xeroxes appeared, only specially authorized, politically reliable people were allowed to use them. This was part of the maniac hysteria of the communist regime, which was afraid of "subversive" propaganda. For the same reason owners of typing machines had to register their machines at the militia and submit to the militia as a sample a page printed on the machine. In this way, if the "class enemy" used the machines to produce "subversive" leaflets, the source of this attack against socialism could be traced down immediately.

Another disciplinary measure against me, which made sensation within the academic life of Romania was that immediately after my application for emigration, the latest issue of the Romanian scientific journal *Studii si Cercetari de Fizica*, in which I had published a paper, was taken out of all bookshops and libraries, and printed again, but this time without my paper. I am not sure whether any other communist country had taken a similar measure against a scientist before or after that. This reminded me of the 1933 Nazi burnings of books of undesired authors, another analogy between the two regimes.

To earn my living I gave private lessons of mathematics and physics to high school students, who were preparing themselves for the university admission examinations. Later I translated physics papers and books from Russian into Romanian. Among others I translated the "Evolution of Physics" by Einstein and Infeld and a popular book on optics by Minnaert. That I had to make the translation from Russian, instead of using the English original, is presumably due to the fact that communist countries, because of lack of hard currency, refused to pay copyrights to Western publishers. Another reason may be that the Russian edition had been already censured and thus was politically more reliable. This applies in particular to the Einstein-Infeld volume since there was a period when some Soviet philosophers had criticized certain aspects of the theory of relativity as being "idealistic" and thus in contradiction with Marxism Leninism.

Being *persona non grata*, the translator's name (my name) was not mentioned in print. An exception to this anonymity rule was a biography on Einstein by the Russian author Lvov, in which it is mentioned that the translation was performed by M. E. Mayer (Hardy) and R. Weiner. I had completely forgotten this book until I found it a few months ago. Hardy does not remember it either, but is pretty sure that he did not contribute to the translation.

He assumes that his name served rather as a fig leave for me: at the time of the publication (1960) he himself was not yet *persona non grata*.

Despite the fact that part of my time was absorbed by translations and private lessons I tried to keep myself informed about the developments in physics and continued to work scientifically, as well. In 1960 Chaim Iussim-Yatom (who had been fired from his assistant professor job at the Physics Department of the University because he had applied for emigration) and I dared even to submit again a paper for publication. We submitted it to *Soviet Physics JETP*, in the assumption that mail to the Soviet Union was less suspicious than mail to the West (Romanian journals were for persons like us a taboo). This assumption turned out to be correct; the paper in which we studied for the first time the influence of the nuclear quadrupole moment on the electron wave function reached the journal and was soon published.[25] This paper had even some resonance.[26] After that, however, Bădărău advised me that our chances of leaving Romania would be impeded by continuing to publish, in particular in a field like nuclear physics. That is one of the reasons why there is a five-year gap in the list of my publications, although I continued to work. Still, I had little contact with people working in my field, either theorists or experimentalists. I found some remedy for my scientific isolation in a kind of private seminar organized in collaboration with some colleagues, who either were in a similar situation, or who were just interested in discussing physics in an informal atmosphere. Among the participants were Josef Ausländer, Erwin Friedlander, Alex Friedmann, Chaim Jussim, Vladimir Rittenberg, and Adam Schwimmer. All these people managed eventually to get out and continued their career as physicists in different countries.

During these difficult years, besides in physics I also found refuge in music. Although I did not play a music instrument — after 1941 and until 1969 I did not have the opportunity even to touch a piano — as a student I went regularly to the musical auditions at the Polytechnical Institute and to the rehearsals of the Bucharest Philharmonic; anyway, tickets for the regular concerts and recitals were quite cheap even for a "retrograded technician." Bucharest was during the late fifties and sixties one of the important music capitals of Europe. This was in part due to the role played by the Enesco festivals, which were actively supported by famous musicians from all over the world. Yehudi Menuhin, a former student of Enesco, and many other celebrated musicians were members of the jury and performed also quite often as soloists in concerts and recitals. I remember an impressive event of this musical activity, when Menuhin had played with the Bucharest Philharmonic the Beethoven concerto and it happened that David Oistrach, whose concert was scheduled for the next day, was in the audience. The enthusiastic public recalled Menuhin several times to the stage and eventually Oistrach joined him and they performed together the concerto for two violins by Bach. I remember this event also because I had not found tickets for this concert, which took place in the Athenaeum Hall. However, in the afternoon before this concert, there was in the same hall a recital by another musician and for this recital I had found a ticket. After that recital I hid myself in the toilets and came out of my hiding just before the Menuhin concert. Given the extraordinary character of this concert the organizers tolerated the standing listeners, among them many professional musicians. On this occasion I had the chance to be able to compare "first hand" these two giants of the violin. It seemed to me that Menuhin's sound was crystal clear, but somehow colder than Oistrach's, which I preferred (both played on Stradivarius violins).

To name just a few other soloists the memorable concerts of which I attended at that time: Arthur Rubinstein, Ida Haendel, Sviatoslav Richter, Emil Gilels, Claudio Arrau, Monique de la Bruchollerie, Annie Fischer, Igor Oistrach. Richter particularly impressed me and I remember a story about one of his concerts in Bucharest. He was scheduled for an evening concert with the Philharmonic. The rehearsal with the orchestra had to take place in the afternoon before the concert, but Richter did not show up.

The conductor (I think it was George Georgescu, the music director of the Philharmonic) sent somebody to look for him at his hotel, the Atheneum Palace, next to the concert hall, but he was not there, nor did his wife know where he was. But then she suddenly realized that in his concert he was scheduled to play one of the piano concerts of Liszt. She immediately guessed that he had gone to the cemetery: After playing a Liszt concert for piano and orchestra Richter used to give as a *bis* Liszt's *Totentanz* and to be in the right mood for this piece he used to go before that to the cemetery. And, indeed, there he was.

The Bucharest Philharmonic was during those years quite a capricious orchestra. Its performances with conductors like Sir John Barbirolli, Carlo Maria Giulini, George Georgescu (their conductor in chief), Georges Prêtre, Constantin Silvestri were in general good. However, if it happened that the conductor was not so illustrious, the musicians, especially the wind players, were unpredictable. The more so I appreciate the quality of French wind players; in my unauthorized opinion the French school of wind players is one of the best in the world.

For many people music is primarily a source of joy and a way to forget the unpleasant things of life. For me, however, it has constituted also a permanent source of inspiration in my work. It seems that while as a listener, you apparently don't "do" anything, the subconscious continues to ruminate and frequently problems, which previously appeared unsolvable; look suddenly in a different light and quite often the solution is obvious. Actually, one of the reasons for our moving to Paris was the richness and accessibility of concerts of classical music.

The Romanian Thaw

In the mid 1960s, in particular after the death of Gheorghiu-Dej, the Secretary General of the Romanian Communist Party, there was a change in the policy of the Romanian government, which became apparent also for outsiders. The relations with the Soviet Union cooled down appreciably and in its relations with other countries, in particular with the West, Romania manifested an increasingly independent point of view. Thus it was the first Communist satellite, which recognized the Federal Republic of Germany and which established diplomatic relations with this country. All the other Communist countries had relations only with communist Eastern Germany. Another and much more dramatic act of independence was the refusal of Romania to join the Soviet Union and the other satellites in the occupation of Czechoslovakia. (As will be pointed out below that this step played a decisive role in my escape to the West).

Although the change of policy of Romania manifested itself mostly in its foreign relations, some collateral effects were also unavoidable in the interior. Thus a slight liberalization of the cultural life took place. In the bookshops and libraries there appeared books by authors like Kafka, which after 1947 had disappeared, because they did not "fit" realistic socialism, plays by Ionesco and Pinter were presented in the theaters, classical music concerts rediscovered modern composers like Schönberg or Berg, and, most importantly, foreign Western newspapers and magazines like *Le Monde* or *Der Spiegel* became available at certain hotels and could also be received by mail from abroad. (The Western newspapers available until then had been limited exclusively to the organs of the Communist Parties, like *L'Humanité* or *Unita*).

Another manifestation of the change of party policy was the fact that travels to other communist countries became easier. While in the period 1959–1964 I had not been allowed to travel even to Bulgaria, in 1965 I could visit Hungary, albeit within a tour organized by the state tourism office. The accompanying Romanian "guide" didn't speak Hungarian or any other language except Romanian, and her duties consisted in supervising closely the group. Nevertheless the train which we took being a night train, I had chosen to take a sleeper compartment, so that I was separated from the group during the journey. To my surprise I shared a sleeping compartment with an American, Joe Policano, who was traveling from Bucharest to Prague. It was a very pleasant surprise, apparently for both of us, since we spent the whole night talking and telling each other the stories of our life. He was by profession journalist and led a public relations company in New York. The purpose of his trip was to investigate, on behalf of American travel companies, the tourism possibilities in Central and Eastern Europe. He appeared surprised about the inefficiency of the tourist organizations in the communist countries he had visited, but otherwise was quite well informed. He had in particular no illusions about the methods and imperialist goals of these countries. As a reporter for the *New York Herald Tribune* in the Korean War he had witnessed the disconcern with which the Chinese commanders sent their soldiers to death; his relating of the waves of thousands and thousands of human bodies trying to roll over the South Korean and American defense lines impressed me, too, although I was no novice in matters of communism, either. Still, for Joe it was a new experience to learn from somebody like me how the communist society, seen from inside, looked like. After I came to the USA we became close friends and our friendship has continued since then and been extended to our families, as well.

Budapest impressed me very positively because it reminded me of Czernowitz before the war, in particular its still Austro-Hungarian atmosphere, the fact that many people spoke German and last but not least through its many cafés (more than 2300 according to the tourist guide). The Hungarians are true coffee addicts. Castro claimed that Hungary had 100 million habitants; he had estimated this number from the quantity of coffee Hungary used to import from Cuba. I met in Budapest during my excursion a former Bucharest colleague of Hungarian origin, who had moved to Hungary some years before. He confirmed what I had already heard: in contrast with the situation in Romania, Hungarians were much freer to travel to the West, either directly or through a detour via Yugoslavia. Another manifestation of the relative freedom

Hungarians enjoyed can be judged from the following: on one of the evenings our group had dinner in a restaurant where a Jazz band was supposed to perform. However on that evening the band was on strike, because the custom office had asked them to pay duty for some musical instruments they had brought from England, where they had been on a concert tour. A strike in a communist country was something unheard of for me. The Hungarian 1956 revolution, although bloodily repressed by the Red Army, had not been for nothing, after all.

We visited also Esztergom, a city in the West of Hungary, where there is the biggest basilica of Hungary. And in this basilica we saw what, according to our Hungarian guide, was the biggest painting of the world. In the museum of Christianity of the basilica, besides many incunabula, I found to my surprise a work by Bourbaki, the famous French mathematics group. Nobody could explain to me what the relation between Christianity and Bourbaki was.

Another collective travel within the Eastern block in which I participated was a cruise on the Black Sea, organized by the Institute of Atomic Physics of the Academy. Although I was not a member of the institute, they accepted me on the list of participants, presumably because they needed a certain minimum number to fill the boat. It was the *Transylvania*, the biggest Romanian cruise ship. To my surprise I shared the cabin with seven passengers, who were all women. Never before or after I had been so spoiled. The boat stopped in three harbors, all in the Soviet Union: Odessa, Yalta, and Sochi. In Odessa I met two of the brothers of my father who had come from Czernowitz to see me. From the city I remember the famous stairs filmed by Eisenstein in his movie *Potemkin* and a big cathedral, which, to my surprise, did not serve anymore as a church, but had been transformed into an office building of the Polytechnic Institute. In Yalta I was impressed by the manor in which the famous Yalta conference took place and the two golden lions in front of this building, which Churchill had wanted to buy from Stalin. Both Yalta and Sochi impressed me among other things through their tropical vegetation; it was for the first time that I saw palm trees.

Interdiction to Leave for the West

The liberalization of travel restrictions did not go so far that I would be allowed to travel on my own to a communist country. As to travel to the West, that remained something beyond dreams, not only for pariahs like myself, but for most citizens. An exception constituted those who were *bought* by relatives or friends. Unfortunately, for reasons, which are unclear to me, this trade did not function in my case. Recently I found out that on the (secret) party committee that decided which scientists were allowed to visit the West, was also Hulubei. Was he responsible for my "privileged" treatment, given the fact that I had been, initially, a member of Andrei Popovici's group, his personal archenemy?

The deal "people for dollars" was usually handled through international commerce organizations with which Romania had trade relations. When importing goods from abroad, the Romanian state, instead of paying in cash, often paid in people. An English businessman very active in this kind of commerce was a Mr. Jacober, through whom thousands of people managed to leave. My relatives in Israel tried to get me out by paying 3000 dollars, the standard market price for a person or a family. For them, recent immigrants, this was quite a big amount of money. However, after several months of negotiating with the Romanian authorities Mr Jacober returned the money because the seller had refused to deliver the merchandize. Whether the amount offered had been too small or there were other reasons behind this refusal, I don't know. The money, however, was eventually lost, because it was deposited at a bank, which soon went broke.

At least two other attempts in this direction, although of different nature, failed. The first attempt was due to the wife of Ionel Perlea, the conductor of the Metropolitan Opera in New York. In its efforts to improve its image in the West, the Romanian Government had started to invite former Romanian citizens, who lived abroad and who had reached world fame, to visit the country. The following joke illustrates the duplicity of this policy. There were three categories of people of Romanian origin living abroad. *Traitors* — meaning those who could not make the ends meet, *Romanian citizens with their domicile abroad* — meaning those who had reached respectable positions and finally, *the great sons of the Romanian people* — meaning those who had gained celebrity. I was befriended in Bucharest with Lucia Cotescu, a stepsister of Mrs. Perlea. (Another stepsister of Lucia and sister of Mrs Perlea was the wife of Mircea Eliade, the famous Romanian philosopher and specialist of religions, at that time professor at the University of Chicago). During the first post-war visit to Romania of Mrs. Perlea, Lucia introduced me to her sister. During her next visit, one or two years later, Mrs. Perlea was received by Gheorghiu-Dej, the secretary general of the Romanian Communist party. On this occasion she handed him over a list of ten people and asked that these people be allowed to leave. On this list was also my name. The result was negative. Sometime later, Bruno Kreisky, the minister of foreign affairs and future chancellor of Austria, visited Romania. He was a friend of Vicky Weisskopf, the director of CERN. Weisskopf knew about me, among other things through another friend of his, Josef Ausländer. Vicky Weisskopf sympathized with my case because he himself had been forced to emigrate and, as he told me when I met him personally for the first time, he had had family in Central Europe, which had a fate similar to mine. Besides that, as somebody who has worked on hyperfine structure and isotopic shift (among other things he published with A. Bohr a paper on this subject) he understood and apparently appreciated my work on the isomeric shift. Kreisky, too, handed over to the Romanian government a list of people for whom he asked exit visa. My name was again on this list, and again without avail.

In the mean time I continued to apply for emigration (in the period 1958–1968 I made 11 official applications), my applications continued to remain unanswered and I continued to be a retrograded "technician." After several years of self-imposed seclusion, I decided to start again to publish, under my private address, in foreign scientific journals, mostly those of the American Physical Society.[27] I guess this was the first time that papers, the authors of which indicated a private address, were published in *Physical Review* and *Physical Review Letters*. I also believe that many readers of these journals did not realize that *Labirint Street, 28, Bucharest* was a private address and not the address of an institute. Indeed, I often got letters and requests of reprints addressed to the "Institute of Physics, Labirint Street 28, Bucharest." For some of these papers I could even order reprints: Hardy Mayer, who was established already in the United States graciously paid for them.

Nuclear Recoil in Muonic Atoms

The isomeric shift had been initially formulated for "normal" atoms, i.e. atoms made of a nucleus and electrons. It did not take long time and the same effect was seen also in muonic atoms. Muonic atoms are analogs of normal atoms where one electron is replaced by a muon. This happens through the capture of a muon by the nucleus. The remarkable feature of muonic atoms is that because of the mass of the muon is two hundred times bigger than that of the electron, the radius of a muonic atom is accordingly two hundred times smaller than that of a normal atom. Therefore the muon is on the average much nearer to the nucleus than an electron and can provide important information about the nucleus.

The role that the muon played in the discovery of parity non-conservation and later also in the isomeric shift, stirred my interest in this particle. In muonic atoms one usually studies the X or gamma rays emitted when the muon bounces from a higher atomic level to a lower one. The nucleus is in this process a spectator, whose structure is not affected by this bounce. It suffers, however a recoil which until the mid-sixties was not considered as relevant. Still, one day I realized that this simple kinematical effect could provide important information about the muonic atom, which could not be obtained by other methods. In a paper published in the *Physical Review*[28] under the title "Nuclear recoil in the decay of bound muons. A possible tool in Muonic-Atom studies" (this was my first paper in the series published with my private address, Labirint 28, Bucharest) I showed that the nuclear recoil following the muon decay in a muonic atom maps almost exactly the wave function of the muon in the atom. By measuring this recoil spectrum one could pin down the atomic state from which the muon decays, in details, which could not be otherwise achieved. I am not aware whether this idea has been so far applied. It might have to await muon factories.

7

Challenging Conventional Wisdom in Particle Physics

It is a fact of scientific life that sensational experimental findings are sometimes not confirmed by subsequent, independent observations. Whether this is due to insufficient statistics, systematic errors or other causes is irrelevant for the following discussion. What I would like to point out here is that even wrong results are often useful for the progress of science, not only because scientists learn from the mistakes of their predecessors, but also because an unexpected result can stimulate the fantasy of the theorist. What is valid for geometry — *the art of deriving correct results from wrong figures* — is often true for physics, too. In my scientific career as a theoretical physicist I was confronted several times with situations of this kind and in the last years of my Bucharest period this happened twice.

Anticipating Electro-Weak Unification?

It is a strange feeling to browse through the elementary particles literature of the 1960s, the period just preceding the standard model. There had been important strides in weak interactions with the discovery that these interactions violate certain symmetries (P, C and PC) considered until then as sacrosanct.[29] Besides that the quark model signaled progress in the domain of strong interactions. Nevertheless a more profound theoretical understanding of these findings was missing. Not only was there no field theory of weak or strong interactions but also many theorists did not believe that field theory was the answer to the unsolved problems of particle physics.[a] Electromagnetic interactions for which there existed a renormalizable, that is consistent, field theory, Quantum electrodynamics (QED), were considered an exception. Not to mention that there was no link between the three interactions characterizing particle physics, except that one suspected — because of very general (dimensional) considerations — that at very high energies the weak interaction would become strong. On the other hand QED had been brilliantly confirmed and resisted all efforts of falsification. The validity of QED used to be parametrized in terms of a length scale Λ of the order of 10^{-15} cm up to which QED had been tested.

And then one day this situation was challenged. In wide-angle photoproduction of electron–positron pairs, deviations from the predictions of conventional QED had apparently been observed.[30] Although eventually not confirmed, this result inspired theorists, me among them, to speculate and to suggest interpretations of this surprising observation. And it happened that the interpretation

[a] My frustration with this unsatisfactory situation went so far that I proposed around that time to McGraw Hill in New York a book dedicated to the comparison of strong, electromagnetic and weak interactions. Although I got a positive reaction from the publisher I eventually did not pursue this project. In hindsight this appears to me a fortunate circumstance, since, had I written this book, it would have been outdated even before its publication: 1967 marked the unification of the electromagnetic and weak interactions.

I suggested, namely that in what was considered at that moment QED charge conjugation C is not conserved, was correct, despite the fact that the result of experiment[30] was not.

The explanation for this curious fact is the following. The interpretation of the results of that experiment relied among other things on charge conjugation invariance C, which had been tested in low energy atomic physics. On the other hand, wide-angle photoproduction was a large momentum transfer process, which essentially meant small distances, below the range where QED had been tested so far. Therefore I wondered whether the above experimental result wasn't an indication that at small distances QED violated charge conjugation. I discussed this possibility in a paper[31] in which I also suggested several experiments to test the symmetries C, P (and also time reversal invariance T) and thus *conventional* QED at large momentum transfer. At that moment nobody knew about the existence of heavy intermediate bosons, responsible for weak interactions, although the idea was floating around. What was even less known was the existence of a heavy *neutral* intermediate boson, leading to neutral currents in weak interactions, although neutral currents have been quite common in electromagnetic processes, being mediated by the exchange of the photon. And it is due to this neutral weak boson that the C symmetryin the electromagnetic interactions experiments I suggested is violated, since the coupling of this boson to electric charges does not conserve C, while the coupling of the photon does. Because of the heaviness of this boson this effect becomes observable only at large values of momentum transfer.

Quite soon Glashow, Salam and Weinberg unified electromagnetic and weak interactions into one renormalisable theory. This theory predicted the existence not only of heavy charged intermediate bosons but, most importantly, also of a neutral heavy boson, and a few years later weak neutral currents were observed, confirming the existence this boson even before its production in the lab. In this way the violation of charge conjugation in electromagnetic processes at large momentum transfer became a direct consequence of the unified electroweak theory.

Anticipating Supersymmetry? Exotic Particles — Bosonic Leptons[33]

Another attempt of mine of challenging conventional particle physics wisdom during the 1960s is that related to exotic particles. Exotic particles in this context are particles the properties of which differ from those observed so far; this refers in particular to particles with unconventional leptonic quantum number and/or unconventional spin, or to leptons with an additional quantum number, strangeness.

The starting inspiration were some experiments which suggested an asymmetry in the angular distribution of muons arising in the decay pion → muon + neutrino. The observed asymmetry in this $\pi \rightarrow \mu + \nu$ decay was at odds with the dominant assumption that the pion had spin zero, but it could not easily reconciled, either, with the assumption that the pion had spin 1. On the other hand it seemed that a particle, which combined the bosonic property of spin with the leptonic property of internal quantum could do the job. Therefore I asked whether there did not exist bosonic leptons after all.[34] In 1967, several years before the formulation of supersymmetry (SUSY), which associates to all observed particles with spin s, partners with spin s-½, this idea was quite exotic. Curiously enough, I found[34] that bosonic leptons with lepton number bigger than 1 could exist in quite an accessible mass range. In particular, if their mass were close to that of pions or K mesons, they would have been confused with these mesons. This mass degeneracy would not only explain why they have not been observed, but it would also explain the controversial pion-muon decay. While the pion-muon decayexperiments were ultimately not confirmed, the idea put forward in Ref. 34 that exotic particles could *hide* themselves behind the masses of well-known particles might still be a fruitful one.

Anticipating Grand Unified Theories? Exotic Particles — Strange Leptons

Within the same context of fundamental, qualitative questions and shortly after I had published the paper on bosonic leptons, I dared to challenge another fundamental assumption, namely that strangeness is limited to hadrons. Starting from the experimental observation that neutral currents are strongly inhibited in semi-leptonic strangeness changing processes of the form $K \to \pi + e^+ + e^-$, I proposed[35] that, in analogy to the inhibition of some neutral currents in pure leptonic processes, the inhibition in semi-leptonic processes is also due to a conservation law. In leptonic processes the fact that the electron and muon have different lepton quantum numbers, and that this quantum number is conserved, explains why a process like muon → electron + photon is forbidden. Similarly I suggested that in semi-leptonic reactions it is the strangeness quantum number, which is conserved, and the variation of strangeness in the hadronic current is compensated by a strange leptonic current. The price to pay for this proposal was to assume the existence of strange leptons. In other words the leptons observed in strangeness changing reactions are different from those observed in strangeness conserving reactions.

This proposal stirred immediately interest and several follow-up papers, among them specially performed experimental studies that appeared in *Physical Review Letters* and in other journals.[36] Most of these papers showed that the existence of strange charged leptons was difficult to reconcile with experimental data and upper limits for strangeness changing semi-leptonic reactions were obtained. One of the most important arguments along these lines[37] had been put forward in my paper already, at the suggestion of Harry Lipkin, who had been the referee of my paper. Contrary to the standard habits of referees who prefer to remain anonymous, Harry Lipkin graciously sent me directly his comments. For strange neutrinos, however, the situation was less clear[38] and I guess that even today this possibility has not been completely ruled out.

With the advent of grand unified theories, GUTs, that is theories, which unify strong and electroweak interactions, this issue got another twist. Indeed, in GUT one expects the existence of new kind of intermediate bosons, lepto-quarks or X-bosons, which carry both lepton and quark flavor quantum numbers and through which leptons can be transformed into quarks and vice-versa. In this sense strange leptons were thus precursors of lepto-quarks.

Around 1967 I got an invitation to CERN for a four-month stay and almost at the same time I was awarded a Humboldt fellowship, which entitled me to spend six months at a German university of my choice and to do research there. Of course, I could not take up those invitations at that time, but they encouraged me to continue to do physics and showed me that my work had not gone unnoticed.

The Escape

It was clear to me that my activity and my correspondence were censored. Although I never had access to scientific or technical secrets, I wasn't sure whether the Securitate — the (in)famous Romanian secret service — knew that. Actually, in the mid sixties a Securitate officer (in civilian clothes) visited me at home, after having introduced himself on the phone as a student who wanted to take private physics lessons. He asked me about my scientific activity and about the papers I was sending abroad. After that I had to report for questioning twice a week to one of the offices of the Securitate (the building had no official identifying plate and nobody would have guessed what was going on there). This questioning continued for several weeks.

I had been aware that the Securitate might consider me a nuclear physicist, of potential value to the "class enemy." Therefore, warned also by Bădărău, I had started to shift the topics of my work from theoretical atomic and nuclear physics to elementary particle physics; the papers quoted above were in part the result of this shift. I thought that particle physics was in the eyes of the Securitate a more "academic" subject. Unfortunately that did not help either.

Around that time I was told (if I remember correctly, again by Bădărău) that if I withdrew my applications, I could obtain my previous position at the institute. It seemed that the publicity my case had gained abroad among physicists and non-physicists had become embarrassing for the government and for the Academy of Sciences. I remember two encounters in Bucharest with foreign physicists. In the early sixties I met Belenkii. I knew him from Dubna and he was familiar with my work. During a visit of his in Bucharest he looked for me. He conveyed me regards from Smorodinskii, in Dubna. People in Dubna had learned about my situation, which even for them, with their Soviet experience, was a "novelty" at that time. They renewed their offer of a position in Dubna and Belenkii, knowing that I had difficulties in making ends meet, offered me financial help. I have been very much impressed by this gesture, which I did not accept. As to the renewal of the Dubna offer, I did not accept that either, because I was afraid that in this way my chances to get out would be even more reduced. In hindsight, had I known that it would take another six or seven years for this to happen, my decision perhaps would have been different. In the late 1960s I also met Walter Thirring in Bucharest, at that time the director of the Theory Division of CERN. I didn't expect that a few years later he would be my host at CERN.

It is worth mentioning that the policy of the Romanian regime towards people who wanted to leave was imitated a few years later by the Soviet Union, where a special name, "refuseniks," was coined for those who refused to renounce their application to leave the country. Sharansky was one of the most famous in that category, although he entered this club later than me. The privilege to have been one of the first refuseniks worldwide is something I could have done without.

At the end of the 1960s the policy of the Romanian government concerning emigration and travel abroad in general started to undergo subtle changes. When you applied for emigration, you had to enclose a statement by your employer that you had not had access to "state secrets" and that your presence at your institution was not indispensable. As a result of this change some applications were, indeed, answered in the positive. Unfortunately, this, again, did not work in my case. Although Bădărău in his quality as director of the Institute gave me such a statement, my application remained unanswered. Bădărău, himself, was a "white" Russian, who had fled to Romania after the revolution and who, after 1945, had not hesitated to turn "red." He repeatedly lectured me about the unpredictability of communist regimes. He used to tell me that it would be a fundamental mistake to assume that the activity of authorities is solely guided by logic.

Formally, it was the Academy, which had to solve my problem. I knew quite well the leading figures in the Mathematics and Physics Section of the Academy; they had been my professors during my university studies and Şerban Ţiţeica, my PhD advisor was now the vice president of the academy. Strangely enough, mathematicians like Miron Nicolescu and Grigore Moisil were much more friendly and forthcoming than the physicists. Ţiţeica, in particular, who also had an important administrative function, was evasive. The reason might be that he was not a party member while Nicolescu and Moisil were and could thus afford more freedom. Eventually the Academy came up with a "solution." At the University of Bucharest a new "Computation Center" had been established and I was offered a position of senior physicist there, which I accepted. The director of the Computation Institute was Grigore Moisil, a very charismatic and witty person (his friends and students called him Grigri) who told me that my case reminded him of similar situations his father was faced with in the late 1930s, as director of the Center of Namismatics of the Romanian Academy. At that time Jews had been ousted from the university because of the anti-Semitic laws of the fascist regime and his father had harbored them in his institute, giving them the possibility to survive. Moisil assured me that in his institute I would be free to continue my research work, because I had been recommended to him by Ţiţeica, whom he had a high esteem for. And what seemed to me quite important — I would not have to withdraw my applications to leave the country. Grigri not only had no illusions about my intentions in this respect but also proved his sympathy for my case in deeds.

1968, the year I joined Moisil's institute, was also the year of invasion of Czechoslovakia by the communist countries led by the Soviet Union. Romania was the only satellite country, which did not participate in the invasion and the process of liberalization concerning travel abroad continued. Among other things, while in the past Romanian citizens had been allowed to travel on their own to other communist countries only for close family visits, now applications for individual tourism were also accepted. For this purpose, however, a recommendation by the director of the institution where one was working had to be enclosed as a sign that you were a reliable "comrade." In this way the right to travel, which had been in the past a privilege of a restricted number of party members, became a reward for "good behavior" for many more people. This change of policy did not apply, however, for travel into capitalist countries, which continued to remain taboo.

To test this liberalization I applied for an exit visa to Bulgaria. While Romania had no frontier with a "capitalist" country, Bulgaria had two such frontiers, with Greece and Turkey. Moreover, I had heard that from Bulgaria excursions were organized to Turkey and that in this way some Romanian citizens had succeeded to escape. Still, my application for Bulgaria was rejected, which proved that even after this "liberalization," some people continued to be less equal than others. There was, nevertheless, a change. In the past, either your applications remained without an answer, or, if you succeeded to be received by somebody, you were given an oral negative answer, and you did not have the possibility to ask why. (When I read for the first time "The Castle" and "The Trial" I had asked myself how Kafka could have foreseen so early — these novels had been written before or during the *first* World War, and were published posthumously in the 1920s — all this). Now in the Ministry of Interior, which handled passport issues, a special service was created, where one could ask for more information about the reasons of such rejections. I used this possibility and was received by a rather high-ranking officer — I think it was a colonel (in all communist countries the Ministry of Interior had military status). I told him the story of my 11 applications for emigration, which, I said, had been motivated by family reasons (this was in part true, my entire family was abroad), and argued that this time I had not asked for emigration, but just for a tourist visa to a "friendly" country. To my surprise my interlocutor told me that he thought this had been a mistake and that I should apply again for one of "these" countries, "these" meaning a socialist country. Given my previous experience, I was very sceptical about the chances of another application, but I had nothing to lose. I applied, therefore, at the beginning of 1969, for a tourist visa to Czechoslovakia. After Bulgaria, I had chosen Czechoslovakia not simply "for a change," but because the "rumor agency" of would-be emigrants had discovered that, although under Soviet occupation, the regime in Czechoslovakia was still quite liberal and that in certain cases Romanian citizens had succeeded to pass to Austria without an exit visa. Actually Moisil, who had given me his OK for this new application, apparently had also heard these rumors: Handing over his recommendation he told me explicitly that he wouldn't mind if I would forget to return. A few days after me, Uli, my future wife, whom I had met in Bucharest some months before, also applied. We had decided to marry, but did not do it before applying for the visa, because we both wanted to leave and we knew that as a married couple our chances to get both the visa were lesser than if we applied separately. And to our surprise, in summer 1969 both Uli and I got the approval and on September 24 we left together for Czechoslovakia. Now the second act of my exit from behind the Iron Curtain had started. While the first one had taken twenty-five years — from the moment my father and I were liberated from the Nazi regime — the second took less than two weeks. My father, unfortunately, did not live to see this moment.

Czechoslovakia

Once arrived in Prague we started to look for a hotel. We had been told that this was not a trivial exercise, as it was tourist high season and the hotels we could afford were much in demand. We had not had time to make hotel reservations in Bucharest because, once we had the much desired exit visa, we took the first available flight. Airline tickets in a socialist society were, like any other goods (except for the "History of the Communist Party of the USSR") scarce. Furthermore, I was not convinced that I had not obtained the visa by mistake and that the authorities might not change their mind; it happened before with other people, who had their passports withdrawn at the last minute.

To our surprise the hotel problem turned out to be in our case, and only in our case, simpler than expected. Indeed, unlike most people who were queuing in front of us at the reception and who were told, after showing their passports, "sorry, no vacancies," we immediately got a room and even a smile from the receptionist. Those who were refused looked at us with envy; they presumably thought that we had reservations. The explanation was another one: they were from other satellite countries (we heard them speaking Polish, Hungarian and German), and next day, when we met Jan Fischer, the Czech theorist whom I knew from Dubna, he told us that we owed this privileged treatment to our Romanian passports. Romania was the only satellite country not involved in the occupation of Czechoslovakia and the Czechs and Slovaks appreciated that. As we were to find out soon, this circumstance also turned out to be decisive in our escape.

The first thing we did in Prague was to call my uncle Muniu in Israel and to tell him that we had managed to get out of Romania.

Neither he nor I had, until the very last minute before our departure, really dared to believe that this would happen. He decided to come immediately to Vienna to wait for us there, in case we succeeded to cross the border. As an Israeli citizen he was not allowed to enter Czechoslovakia, because Czechoslovakia, like all other Communist countries (with the exception of Romania) had severed, after the Six Day War of 1967, diplomatic relations with Israel. Then we went to a travel office to inquire about the possibility of going to Austria. Austria was the only non-communist country to which Romanian citizen could travel without an (Austrian) *entry* visa. However this did not help, because the travel office confirmed what we already had heard in Romania: to leave Czechoslovakia for Austria we also needed a Romanian *exit* visa, specifically valid for Austria; this, of course, we did not have. Still, we learned something new: There were four possibilities of going to Austria: by plane (from Prague), by train, by boat (on the Danube, from Bratislava) or by taxi (from Bratislava, which was only an hour from Vienna), and, when using the train, the boat or a taxi, passports and exit visas were checked only at the border. And there it was somehow a question of luck. Furthermore, tickets to Austria had to be paid in Western currency and, what was for us a big surprise, in Czechoslovakia you were allowed to *possess* Western currency. In Romania this was not permitted; possession of Western currency was punished with several years imprisonment. Since a flight Prague-Vienna was out of question because one asked for the passport at check-in, we concluded that Bratislava appeared the most promising avenue for our intentions. For Jan Fisher, whom we informed about our plans and asked for advice, our problem was something completely new, because, despite the occupation, Czechoslovak citizens could travel freely. The "Prague-Spring" had still left some traces. As a matter of fact, Czechoslovakia appeared to be also in other respects, e.g. with regards to consumer goods, in a much better situation than Romania. While in Bucharest meat and meat products of any type were rare, in Prague the windows of butchers were overflowed with such products, although people were complaining that you did not find beef, *and* veal *and* pork everyday. The scarcity of consumer goods in a socialist economy is best described by the following joke: A customer enters a sausage shop. He finds two vendors who are reading the newspaper. The customer dares to disturb them and asks for ham. "Sorry, we are out of ham." "What about bacon?" "Sorry, we are out of bacon." "Have you got perhaps Mortadella, Parisian sausage, any kind of sausages?" "Sorry, no sausages, either." After the disappointed customer left the shop, one vendor says to the other: "What memory!"

Jan Fisher had no idea about what was going on at the border with people in our situation, but he told us that the coming week the regular annual triangle physics meeting Vienna-Budapest-Bratislava would take place in Bratislava, and that Jan Pisut, from the University of Bratislava, was the organizer. Fisher advised me to contact Pisut who, as a local of Bratislava, might know more about what interested us. I did not know Pisut, but when I called him up he immediately invited me cordially to the meeting and booked a room for us in a pension in Bratislava, at the expense of the Bratislava University. The next thing we did, still in Prague, was to call a good friend of mine in Geneva, Rachel (Kati) Szekely-Faltin, and ask her to send us some hard currency, so we would be able to buy tickets for Austria. Since we did not have a bank account in Czechoslovakia we asked her to send us the currency by money order to the pension in Bratislava where we were supposed to stay. Kati, however, had a better idea. She contacted Walter Thirring at CERN. Kati, who was an opera singer, knew Thirring very well; as a passionate musician he used to accompany her on the piano in Lieder concerts. Thirring, being from Vienna, knew about the triangle meeting. He called up his colleague Herbert Pietschmann at the Institute of Theoretical Physics in Vienna. Pietschmann planned to participate in the triangle meeting and Thirring asked him to transfer me in Bratislava the money sent by Kati. In this way, with the help of the "international physics mafia," we had solved the money problem, albeit it turned out that we wouldn't really need it, because of another lucky circumstance.

Once in Bratislava we continued our travel investigations. The boat possibility had to be ruled out, because, contrary to what we were told in Prague, it appeared that passports were checked before boarding the boat, after all. The taxi possibility, however, seemed quite doable. We found a taxi driver who was willing to drive us to Vienna. He did not care about our visa problem, "this is your problem." Should the border authorities not allow us to cross, he would simply drive us back. This was for us a surprising statement, which we did not completely trust. The point was that according to Romanian law, any attempt to cross the border without legal documents was a crime, punishable with prison. Our doubts were enforced by the following circumstance. Since the border was not very far from Bratislava, we decided to take a walk through the woods in the direction of the border, to get a "feeling" for what was going on. We arrived at a barbed-wire fence with martial posters in Slovak and Russian, forbidding the crossing under death penalty. For those who did not know Slovak or Russian, the sketches of dead sculls and of bones explained the text. On our way we met an old man who, after he heard us speak a foreign language, guessed what were after. He told us that the border was guarded by Soviet soldiers, who shot without hesitation anybody trying to cross.

Next day, however, a new event took place. While we were trying to cross the border from Czechoslovakia to Austria, my uncle, who had just arrived in Vienna, was trying to cross the border in the opposite direction. And, to his and our surprise, he managed to obtain at the Czechoslovak consulate in Vienna an entry visa to Czechoslovakia, valid 24 hours. He arrived on the same day, by train, in Bratislava. And he had important news: at the Israeli embassy in Vienna he had met several people from Romania who had recently managed to get out of Czechoslovakia. And, what seemed for us even more important, before succeeding some of them had tried several times to cross the border, mostly by car, and were refused. But — and this was for us the essential information — they had just been told "sorry" and could return, without being arrested and handed over to the Romanian authorities, as required by law. This confirmed what we had heard in Prague and from the Bratislava taxi driver and encouraged by this news, we decided to try our luck by taxi and we ordered the taxi for the next morning. In the mean time, since my uncle had to return to Vienna, we accompanied him to the Bratislava railway station. Here we saw many Bratislava people waiting for the train to Vienna. It was a Saturday and they intended to spend the weekend in Vienna. They did not have our visa problems; their Czechoslovak ID cards were sufficient. Five minutes before the departure of the train — we had already made the departure photo (cf. photo 6) — Uli came up with the proposal: "Why don't we take the train just to the border to see how the formalities proceed there?"

Indeed, since one could buy at the Bratislava station tickets up to the border this seemed to be quite an inoffensive act, the more so that you were not asked at the ticket counter to show a passport. We decided on the spot to do that and boarded the train together with my uncle. In about twenty minutes we arrived at the border, where the train stopped. We waited to see what would happen. In case we were asked what we were doing in the train without valid travel documents, we intended to explain that we wanted to spend the weekend in Vienna. We had, of course, no luggage with us and hoped that this "proved" our intention to return to Bratislava. Actually, we expected to be asked to leave the train and to take the next train back to Bratislava.

None of these things happened. An old Slovak custom officer entered our compartment and asked for the passports. My uncle handed over his Israeli passport; the officer checked it briefly and stamped it. Next it was my turn. The customer browsed through my passport looking, apparently, for the exit visa, which, obviously, he did not find. He checked twice but eventually, without asking any questions and without saying a single word, he took out of his pocket his rubber stamp and stamped my passport, as well. The same thing happened with Uli's passport. In Uli's case the browsing took a bit more, (we later speculated that this was due to her maiden name, Culic, which had a Russian twist and for which the old man might have had less sympathy), but eventually he stamped it, too. He left the compartment and soon the train moved. We looked one to another and did not know what was happening. But suddenly we realized that we had crossed the border into Austria, because an Austrian conductor entered the compartment and asked in German for the tickets. My uncle had a return ticket, we of course not. But that was no problem: My uncle bought for us tickets from the conductor, with Austrian Schillings — the dollars that Herbert Pietschmann had brought us were not necessary for the moment — and in half an hour we were in Vienna. And we were free!

8

Nazi-Communist Analogy

The similarities between Nazism and Communism have been the subject of many scientific studies. Among these the work of Hannah Arendt and in particular her classical book "The Origins of Totalitarism"[39] is considered by social scientists as the cornerstone of the political history of Europe in the 20th century. I have read this book quite recently and was impressed by its richness of information, sharpness of analysis and stylistic beauty and I recommend it without hesitation to those interested in the subject.

Having spent my youth under both the Nazi and Communist regimes, I am tempted to draw an analogy between these two dictatorships from a purely subjective point of view, from the point of view of a witness albeit one who had been in both cases on the "wrong" side.

Both regimes violated the fundamental human rights, the freedom of movement, speech and religion, of want and fear.

Both dictatorships pretended to serve a certain category of people: under fascism and Nazism it was the race, under communism the working class.

In reality, in both cases not only was this claim not substantiated but it also led to unprecedented sufferings and hardships of those who should have benefited from it. Never before had the German people undergone bigger trials and tribulations than those, which were imposed upon it by the war and after-war events. This includes the sufferings of the civil population during the bombardments, the displacement of millions of people from East to West, and the division of Germany. Never before had the working and in particular the peasant classes of Eastern Europe and Asia (that constituted essentially the vast majority of the populations of the Soviet Union and its European satellites, of China, Vietnam, Cambodia and North Korea) which should have been the beneficiaries of the Communist revolution, suffered more than under this Communist rule.

Both regimes were imperialistic and both regimes were after world domination.

Both regimes were based on personality cults and the leaders of both regimes were criminals.

Both dictatorships had tried to base their rule on an ideology and in both cases this ideology was fundamentally flawed. This ideology was propagated in the most abusive way: no divergent opinion was tolerated and the press and radio (in the communist case, which lasted longer, this applied also to television) were under strict state control. The disinformation of the public opinion started in school and applied to all cultural activities. History was falsified and rewritten every day, in the letter and spirit of Orwell's "1984."

The economies of both regimes were from the beginning condemned to failure. The Nazi economy could survive only by starting World War II, conquering the rest of Europe and enslaving and exploiting other people, while the communist economies survived by enslaving and exploiting their own people.

Both the Communist and the Nazi regimes used concentration camps (in Germany these were then transformed into extermination camps) and in both regimes forced labor was an important economic factor.

Both regimes were anti-Semitic.

There were of course many important differences between the two evils, which are, sometimes, overlooked, in particular by those who suffered the tragic consequences of only one of the two. In Germany this tendency of overlooking has been, until recently, particularly strong. I will mention below some of these differences, as they appear from my own perspective.

Nazism was a popular phenomenon, supported initially by the majority of the German people. Actually, Nazism was established in Germany by democratic vote and, until the war turned out to be lost, most Germans were happy with the regime.[40] Communism, on the other hand, conquered the power in Russia through a revolution lead by a minority and in Eastern and Central Europe through sheer military occupation by the Red Army. And, unlike the Nazi regime, it was never popular.

The communist ideology, although not applicable in true life, because it ignores certain fundamental facts, has at least the merit of being utopian and thus morally justified. After all, the idea of eliminating the exploitation of men by men, of making available to everybody everything an individual needs, is certainly not condemnable. The main flaw of this idea is that it just does not work. Still, the difference between "The Capital" of Marx and "Mein Kampf" of Hitler needs no further comment. And, with the risk of being cynical and overlooking the specific geopolitical differences when comparing dictators it must be acknowledged that Stalin was a much more "gifted" dictator than Hitler: his regime (and that of his successors) survived more than seventy years, while Hitler's only twelve.

On the other hand the Nazi ideology propagating the dominance of one race over other races had from the very beginning no justification, whatsoever, either moral or scientific. As a matter of fact, what the Nazi ideology considered the most inferior of races, the Jewish race, is, from the scientific point of view, not a race. Furthermore, there can be no comparison between the crimes committed by the Nazis and those by the communists: the Holocaust (but not the genocide[a]) is something unique in history, because never before had the destruction of a people been organized on an industrial basis. The Nazi regime intended to, and essentially succeeded, in *exterminating* the entire Jewish population in that part of Europe, which it controlled.

As to the communist anti-Semitism, it was not ideological, but rather opportunistic. Any dictatorship needs a scapegoat to divert the attention of the masses from its wrongdoings and the communist one was no exception to this rule. Jews had served for two thousand years all over Europe as convenient scapegoats and they did this job also in the Soviet Union and its satellites. Officially no communist regime has ever pursued an anti-Semite policy. In practice, however, under the cover of anti-religious or, later, anti-Zionist campaigns, this turned out be the case. For the communist regimes Zionism was considered a capitalistic ideology and as such strongly condemned. In the Soviet Union this policy made part of the opportunistic and contradictory policy of Stalin towards national and religious minorities. Jews were particularly targeted, among other things also because of the failure of the experiment to settle them in an autonomous region, Birobidzhan, in the Far East of the Soviet Union. After the war not only were, in the USSR, manifestations of Jewish culture like Yiddish theaters and newspapers forbidden, but Jews were sometimes singled out as culprits in show trials, like Stalin's doctors plot. Even after Stalin's death, under the pretext of a national minorities policy, Jews were discriminated.[41] Thus, while in the first days of the revolution there were many Jews in high-ranking state and party positions, in the post-war era there were almost none. Jews had also great difficulties entering universities: Khrushchev was quoted as saying "as many Jews work in the mines, as many should be in the universities." The attitude of the communist regimes towards Israel was also, at a first look, contradictory, but in reality

[a] The only communist regime, which practiced on such a scale genocide, albeit not by industrial means, was probably the Pol Pot regime in Cambodia.

it was always dictated by concrete political interests. In 1947 the Soviet Union played a key role in the United Nations' decision to divide the former territory of Palestine into a Jewish and Arab state and it was among the first countries to recognize Israel. This decision was not inspired by sympathy for Zionism and the fate of Jewish survivors of the holocaust, but by the fact that it constituted a blow against the British Empire, which had kept until then the mandate for Palestine. Very soon this attitude would change and after the 1967 Israeli-Arab war the Soviet Union and most of its satellites severed even their diplomatic relations with the state of Israel. The USSR and the other communist countries became the main supporters and arms suppliers of the Arab neighbors of Israel, and anti-Zionism and anti-Israel policy became an official component of communist policy. This change of foreign policy, which was followed by all satellites, manifested itself, among other things, by often denying the survivors of the holocaust the right to emigrate to Israel and sometimes even in show trials: one of the most famous ones in this category was that of Rudolf Slansky, the secretary general of the Czechoslovak Communist party, who was sentenced to death as a Zionist agent.

However the Soviet anti-Semitism was, with certain exceptions, less brutal and less criminal than the fascist one; after all, most Jews who fled from the Nazis to the Soviet Union and reached the Eastern regions of the USSR, which were not occupied by the German army, as well as many of those who were deported in 1941 to Siberia, because of their bourgeois origin, survived.

Section III

Geneva, Bonn; Statistical Concepts in High-Energy Physics (1969–1974)

9

CERN

From Vienna to Geneva

We spent about ten days in Vienna waiting for the entry visa to Switzerland. For me, as an invitee of CERN, the Swiss entry visa formalities were quite simple; for Uli it was somewhat more complicated, because we were not yet married. We tried to marry in Vienna, but the formalities would have taken three weeks. However CERN managed to handle this unusual situation by arguing that Uli was my fiancée; as a matter of fact, her Swiss entry visa, which she got within a week, specified explicitly that she was accompanying me, her fiancé. My CERN colleagues told me later that this was some kind of premiere for the Swiss authorities. The tragic case of my countryman Joseph Schmidt, who, during World War II, had been denied refuge by the Swiss authorities, and that of other thousands of fugitives of the Nazi regime was (and is) still vivid in my mind. But times had changed. Although the Swiss knew that we did not intend to return to Romania, they granted us the entry visas. Moreover, a short time after my arrival in Switzerland I was called by the Geneva canton authorities and was offered political asylum. For reasons to be explained below, I did not make use of this offer, but I continue to be grateful to Switzerland for this gesture.

Although Vienna was not anymore *the* Vienna about which my parents and grand parents had told me so many interesting things, the waiting was a very pleasant interlude. For the first time in our adult life we were free persons, and I tried to imagine, sometimes with success, what the life in Vienna before World War I, to which the memories of my parents referred to looked like. The Opera and theaters were too expensive for us, so we were content with the monuments, museums, churches, and coffee shops. Some of the famous *Kaffeehäuser* like *Central*, *Demel* and *Sacher* still existed, but the people who had made them famous had disappeared. At a first look Vienna gave us the impression of an empire capital, which had lost its empire. Still, this impression was to be qualified.

During this short stay I was invited to give a colloquium at the Institute for Theoretical Physics of Vienna University. Then we went for lunch to a restaurant. After lunch Herbert Pietschmann, my host, made a point in trying to convince me that the *Viennese atmosphere* was still alive. Indeed, instead of taking the coffee in the same restaurant, we went for coffee at a Kaffeehaus nearby. According to the Viennese tradition it would have been a sacrilege to drink coffee in the restaurant where you had lunch. Not in vain a leading Austrian psychologist, Adolf Storfer once said: "Austria is not geographical notion, but a state of mind." My colleagues in Vienna, most of whom I had never met before, contributed also in other ways to make our wait in Vienna enjoyable, for example by inviting us to their homes. I had the impression that they treated me, because of my Czernowitz origin, as a countryman of theirs. The old Austrian-Hungarian empire still seemed to exist, after all, at least in the minds of some Austrians. That impression was confirmed a few years later when I was invited to attend another Bratislava-Budapest-Vienna triangle conference, which took place in Vienna. This time I was not just passing through Vienna, as it happened in 1969, but I was spending the academic 1972–1973 year at Imperial College in London. Still, this did not deter my Viennese colleagues to pay my travel expenses and I was pleasantly surprised to meet with the occasion of this conference two other former Austrians, Paul Kessler and Fritz Rohrlich. Both had been forced by the Nazis to emigrate. Fritz Rohrlich told me that he even considered himself in some sense Bukowinian, because his parents were from Suceava.

CERN

We went to Geneva by train where we were waited on at the railway station by Eugenie Iseli, a collaborator of the Scientific Documentation Service of CERN. She was an acquaintance I had never met personally before; she had been sending me preprints and various documentation material to my Bucharest "Institute" at Labirint 28. We became friends and even now, after so many years, whenever we come to Geneva, we call her up and meet her. Geneva, although much smaller than Vienna, impressed us by its international character. Besides CERN, the European Institute for Nuclear Research, with its 6000 employees probably the biggest research institute in the world, Geneva accommodated many other international organizations like the International Red Cross, the World Health Organization WHO, the International Labor Organization ILO, and, of course, the European headquarters of the United Nations Organization. Actually, while after World War II the head office of UNO was established in New York, its predecessor, the League of Nations had its world headquarters in Geneva.

Geneva's international character is due in part to its geography, being at the border between Switzerland and France. CERN itself extends both on Swiss and French territory. There are CERN buildings and laboratories that can be reached from the main administrative center (where also the theoretical physics division is located) by car only by crossing several times the winding border. While cars crossing the border are stopped by the customs officials, pedestrians usually are not. This sensation of being able to cross freely the border was for us the most tangible confirmation that we were in a free country. Actually, with a Romanian passport we needed in principle also a French entry visa to cross the border to France, but nobody cared about pedestrians. In 1969, when we arrived for the first time in Geneva, and even several years later we practiced this elementary right of free movement, which we were deprived of practically our whole previous life, as often as possible. Related to this border story is another observation, which for us who were coming "from the cold" appeared quite amusing. If you were in a car with a Western license plate, at the entry into Switzerland the Swiss border guard did not ask for identification documents: He asked only whether you had something to declare. But this question was followed up quite meticulously; Geneva residents were not supposed to import duty-free more than 200 g of butter, a pound of meat, one bottle of wine, etc. Similar restrictions apply even today. On the other side, however, when entering France the French custom officials did not care to even show up.

Geneva also impressed us by its natural beauty. Even now, after having traveled a lot I continue to consider it, at least from the geographic point of view, one of the most beautiful cities of the world. At the foot of the Alps, with the Mont Blanc and the Jura mountains in the background and its magnificent lake, it offers almost everything a tourist can ask for.

At CERN I felt myself like a fish in water from the very beginning. CERN was the most important international crossroad of high-energy physics, and for a physicist the possibility to work there meant, among other things, meeting and discussing physics with the most qualified people in your field, from all over the world. Besides that, for a phenomenological theoretical physicist like myself who was interested in new effects, there was the unique possibility to meet on the spot experimentalists and to check whether your ideas were realistic or not, whether a certain experiment was doable or not. Anybody engaged in research knows that the separation of the wheat from the chaff is one of the most important steps in the complex process of exploring the unknown, and for this purpose CERN was a wonderful place. In hindsight, however, it seems to me that experimentalists at CERN were a bit conservative. This may be due in part to the fact that running time at a CERN accelerator is very precious: In the last thirty years CERN accelerators had been almost unique tools and the competition was accordingly severe. It may be also due to the difference in mentality between the Old World of Europe, where, when you came up with a new proposal, you were faced with the question "why," and the New World, where one would rather ask "why not."

Not only I enjoyed my stay at CERN, but Uli too. She, by training an oil engineer, joined the chemical lab of CERN and this turned out to be a useful experience not only for her: when we left Geneva the leader of her lab Borghini was also very appreciative of her contributions.

Although we arrived in Geneva essentially without anything — we had bought, upon arrival in Vienna, the most elementary things like toilet things, pajamas, shirts and underwear (when we left Bratislava it was raining; therefore we had umbrellas with us, but, we did not even have toothbrushes) and our future was not determined for the coming 10 months (the duration of the stay at CERN and the following Humboldt fellowship) — we enjoyed our stay from the start. Actually, a few weeks after our arrival, we had a surprise. Matts Roos from Helsinki, whom I had not met before, called me up in my office and told me that he was at CERN and that he had in his car the luggage we had left in the pension in Bratislava. On his way from Helsinki he had passed through Bratislava. Here Jan Pisut, whom I had called from Vienna to ask him to tell the landlady that we would not return, handed him over the few things we had left these. We wouldn't have dreamed to see these things again and they were, at least in the beginning, quite useful.

Uproar in the Media

On October 23, I had another surprise of quite different nature, although also related to our escape. A journalist of the *Daily Express*, the London newspaper, called me up in my office and asked for an urgent interview. He told me that he knew that I have just crossed the iron curtain and that I didn't intend to go back to Romania. He wanted details about my escape and about my reasons for leaving Romania; he appeared, though, to know that I had been persona non grata of the communist regime. The issue was so urgent for him that he wanted to join me immediately.

I was reluctant to meet him because Uli's mother was still in Romania and I was afraid that headlines about our escape might provoke the Romanian authorities to take revenge on her and to use her as a pledge to blackmail us. Precedents of this kind were current practice in communist and fascist countries. To gain time I told him that I had other commitments on that day and we agreed to meet the next day at 10 a.m. in my office. After that telephone conversation I asked my CERN colleagues for advice. As far as I remember, Walter Thirring, the director of the Theory Division, was not in Geneva, so I went to Freddie Herz, at that time the leader of an experimental group at CERN whom I knew quite well through Erwin Friedlander, with whom he collaborated. Freddie, of Austrian origin and also a holocaust survivor, but who had lived in the West since World War II, had more experience with Western media. He not only understood my hesitation to meet the reporter, but it turned out that CERN, too, would have preferred to avoid publicity of this kind. Indeed, Freddie took me to Owen Lock, the "Foreign Minister" of CERN, who was in charge of the international scientific collaboration program; within half an hour we ended up in Bernard Gregory's office, the Director General of CERN. Gregory explained to me that CERN was not happy about this story, because it might jeopardize the collaboration with the scientific institutes in Eastern Europe, which constituted an important goal of CERN; that this collaboration was the only opportunity for physicists from the other side of the curtain to visit CERN and to work here for some time. He explained that although I had not come to CERN within that collaboration program (how could I? my "Institute" — Labirint 28, Bucharest — was not on the list of Romanian Institutes) but as an invitee of CERN, my presence at CERN could be used by cold warriors in the East to claim that CERN accommodated refugee-physicists. That, although it was up to me to decide whether I should give the interview, CERN would favor another approach.

Gregory, of course, forced open doors on me and we agreed that next morning I would not come to CERN and that Gregory would receive the reporter in my place. He would tell him that I had to leave Geneva on an urgent mission. To make sure that the reporter wouldn't find me at home, Freddie and Owen picked me up and drove me to the Geneva airport, where in the airport restaurant we spent the whole morning.

Gregory's strategy worked, although not exactly as planned. I learned about what happened in the evening of that day from the broadcast of *Free Europe* (I still had the habit to listen, almost daily, to these emissions) and next day from Gregory, personally. At 10 a.m. of that day not only did the reporter of *Daily Express* show up at CERN, but more than twenty other journalists of all major newspapers and broadcast stations accredited at the United Nations in Geneva. And they became pretty mad when they learned that they had to content themselves with Gregory rather than with me, and did not hide their dissatisfaction in the choice of words addressed to Gregory. What was, however, more important, on that evening *Radio Free Europe* brought a long report about myself, about my adventurous escape and about the CERN incident. Among other things, the report presented me as a "leading atomic physicist" who had been for many years prevented from leaving Romania. Many, but not all, details were exact. Thus, for example, the report mentioned that I had arrived in Geneva with my wife. This technical error was, from my point of view, welcomed, because were the family name of Uli mentioned, this could have had unpleasant consequences for her mother. Other details, like the trajectory of my flight, were correct. Strangely enough, the Romanian authorities reacted quite slowly; even weeks later Alex Friedman succeeded in getting out on the same route, albeit not by train but by taking the shuttle boat between Bratislava and Vienna, and despite the fact that passports were checked upon embarkation.

On the morning of the next day, on October 24, many newspapers brought the news of my escape. I myself found it in the *Journal de Genève*. Friends from Germany sent me newspaper clippings from different German newspapers, among others also the prestigious *Süddeutsche Zeitung* and *Hannoversche Allgemeine Zeitung*.

Hannoversche Allgemeine Zeitung, Oct. 24, 1969
Romanian atomic scientist escapes in the West

From our correspondent

Geneva, October 23

The defection, kept so far under strict secrecy, of a leading Romanian atomic scientist, has created some problems for the European Organization of Atomic Research (CERN) in Meyrin, near Geneva, which might have repercussions for the future collaboration with research institutions of the Eastern Bloc.

The expert who escaped to the West is the 36 year-old (I was actually 39) atomic theorist Richard Weiner. Accompanied by his wife he had organized his flight, which led them from Bucharest via Prague and Bratislava (Preßburg) to Vienna. In the Austrian capital he obtained from the Swiss embassy an entry visa and has been immediately employed by CERN.

It seems, however, that this research organization — the German Federal Republic is one of its twelve member states — has been afraid of diplomatic complications and therefore took pains to keep secret the presence of the expert. It has been confirmed that, after concluding his work in Geneva, the Romanian scientist intends to continue his activity in the German Federal Republic, where another research fellowship awaits him. Romania had always forbidden him leaving the country.

CERN experts declared in private that Weiner, despite his youth, is an outstanding specialist and theorist, in particular in the domain of high energies. His qualities are estimated so highly that a CERN scientist even stated: "Because of his age Weiner has not received the Nobel Prize yet... one only has to wait a bit."

Rumänischer Atomforscher setzte sich in den Westen ab

Von unserem Korrespondenten

hö. **Genf,** 23. Oktober

Der bisher streng geheimgehaltene Absprung eines führenden rumänischen Atomwissenschaftlers hat der Europäischen Atomforschungsorganisation (Cern) in Meyrin bei Genf einige Probleme gebracht, die sich sogar auf die künftige Zusammenarbeit mit Forschungsstellen im Ostblock auswirken können.

Bei dem in den Westen geflüchteten Experten handelt es sich um den 36jährigen Atomtheoretiker Richard Weiner. Er hatte in Begleitung seiner Frau die Flucht organisiert, die beide von Bukarest über Prag und schließlich Bratislava (Preßburg) nach Wien führte. In der österreichischen Hauptstadt erhielt er von der Schweizer Botschaft ein Einreisevisum und wurde am Genfer Sitz des Cern sofort eingestellt.

Anscheinend hatte man aber in der Forschungsorganisation, unter deren zwölf Mitgliedern sich auch die Bundesrepublik befindet, diplomatische Verwicklungen befürchtet und sich deswegen sehr bemüht, die Anwesenheit des Experten zu verheimlichen. Wie festgestellt wird, beabsichtigt der rumänische Wissenschaftler nach Abschluß seiner Arbeiten in Genf, seine Tätigkeit in der Bundesrepublik fortzusetzen, wo ihn bereits ein weiteres Forschungsstipendium erwartet. Rumänien hatte dem Wissenschaftler stets die Ausreise verboten.

Wie Cern-Experten privat erklären, handelt es sich bei Weiner trotz dessen Jugend um einen hervorragenden Fachmann und Theoretiker besonders auf dem Gebiet für hohe Energien. Seine Qualitäten werden derart hoch eingeschätzt, daß ein Cern-Wissenschaftler sogar erklärte: „Wegen seines Alters kann Weiner den Nobelpreis noch nicht bekommen . . . Man muß nur noch einige Zeit warten."

Kabinett will heute über Aufwertung entscheiden

Von unserem Korrespondenten

ze. **Bonn,** 23. Oktober

Voraussichtlich schon zum Wochenende wird die neue Bundesregierung eine formelle Aufwertung der D-Mark beschließen und damit zum System fester Wechselkurse zurückkehren. Eine Einigung über die Höhe des Aufwertungssatzes und das Problem des Einkommensausgleichs für die deutschen Bauern wird auf der für Freitag angesetzten zweiten Kabinettsitzung erwartet. Die EWG-Kommission und der EWG-Währungsausschuß sollen statt am Sonnabend bereits am Freitag konsultiert werden. Damit würde einer Aufwertung der D-Mark vor Eröffnung der Börsen am Montag nichts mehr entgegenstehen. Nach wie vor gilt ein Aufwertungssatz von 8,1 Prozent – das entspricht einem Dollarkurs von 3,70 DM – als wahrscheinlich.

(Siehe auch Wirtschaft)

Nobelpreis an Samuel Beckett

dpa. **Stockholm,** 23. Oktober

Der irische Schriftsteller Samuel Beckett erhält den Nobelpreis für Literatur. Der Preis, den die schwedische Akademie am Donnerstag verliehen hat, ist mit 290 000 DM dotiert und soll dem Schriftsteller am 10. Dezember in Stockholm überreicht werden. In der Begründung für die Auszeichnung heißt es, Beckett erhalte den Preis für „eine Dichtung, die in neuen Formen des Romans und des Dramas aus der Verlassenheit des modernen Menschen ihre künstlerische Überhöhung erreicht". Zu den bekanntesten Werken des 63 Jahre alten Schriftstellers gehören „Warten auf Godot", „Endspiel" und „Das letzte Band".

(Siehe auch Feuilleton)

To even further increase the thrill, the same page brought the news about Samuel Beckett receiving the 1969 Nobel Prize for literature. Moreover, three days later I got a letter from the Director of the Romanian Department of Radio Free Europe, Noel Bernard, inviting me to Munich, the headquarters of the Radio. Here is the translation of the letter from Romanian.

Monsieur Richard Wiener[42] *27.10.1969*
Centre Européen pour les
Recherches Nucléaires
Genève

Dear Mr. Weiner,

I have learnt from our Geneva correspondent, Mrs. Nicolette Frank that you have remained in the West. I am also informed that C.E.R.N does not want to make noise about your case and has discouraged your contacts with the press. Nevertheless, I would like very much to meet you and assure you that nothing of what we will discuss in our meeting will be used in the broadcasts of our radio station. Should you agree, please let me know.

Yours sincerely,

Noel Bernard
Director
Romanian Department

For the reasons mentioned above I did not take up this invitation either. Up to the present I don't know where the media got the news and all the details of our escape from. Also surprising was that this development took place exactly one month after our escape. A possible explanation of this month delay is that our Romanian exit visa expired on October 24 that is by that date we were supposed to return. Somebody who knew that and who knew that we were still in Geneva might have concluded that we did not intend to go back. Soon after that, while preparing the marriage formalities (we were married in Geneva on December 19, 1969) we got the political asylum offer mentioned above from the Geneva authorities. This gesture from a country famous for its restrictive immigration policy impressed not only us, but also our friends and colleagues. Nevertheless — mainly for professional reasons — we did not use this opportunity. I wanted to continue to work in high-energy physics. It would have taken at least twelve years until we could get

Swiss citizenship. At CERN I could not get a permanent position, if not for another reason but because only citizens of the member countries were entitled to that. Even if I had found a job in my field at a Swiss university, traveling abroad, which in high-energy physics is a must, would have been quite difficult; as a stateless person, I would have needed entry visa for all other countries. Actually, in 1970 or 1971 I was considered for a professorship at the University of Zürich and invited for a colloquium to Zürich, but the invitation reached me with a delay of several months, because I was at that time already in the USA. On the other hand I knew, or, as will be explained later on, thought to know, that because of my German background I could get citizenship immediately in Germany, which would solve at least the traveling problem. Furthermore since I had the Humboldt fellowship, which I could take up any time, the first six months of my stay in Germany were assured. Therefore soon after our arrival in Geneva we decided to move, at the end of my CERN appointment, to Germany. As a matter of fact, soon after my arrival in Geneva, I got also from Yuval Neeman an offer to join the University of Tel Aviv as a senior lecturer. Neeman, at that time the chairman of the Physics Department in Tel Aviv, had learned through Harry Lipkin, whom I contacted from Geneva, about my escape. While still in Romania I had had with Harry Lipkin an exchange of correspondence on my *Physical Review Letters* paper about "Strange Leptons" and other physics-related issues of common interest. Although previously I had not been in either of the two countries, my decision to go to Germany rather than to Israel was determined by several factors. Despite my past, I thought that at least linguistically I would feel more at home in Germany. I also had some health problems (a heart condition, a heritage of the ghetto) and I was afraid I might not stand the Israel climate. Furthermore, I knew that the offer to join the University of Tel Aviv would be standing also after accepting my Humboldt award. Last but not least I should mention that Magda Ericson, who had a joint appointment at CERN and at the University of Lyon, asked me whether I would like to join the nuclear physics group in Lyon. I think I didn't accept this offer because of the citizenship issue, as well as because at that time I wasn't interested in nuclear physics.

Physics at CERN

In 1969, the year of my arrival at CERN, I was interested in strong interactions and in particular in the quark model (I had just published a paper in *Nuclear Physics* on that subject). However I also followed the developments in weak and electromagnetic interactions and had long discussions on these interactions and on more general issues with people from or visiting CERN, among others Daniele Amati, John Bell, Andrei Bialas, Giuseppe Cocconi, Magda and Torleif Ericson, Sergio Fubini, Rolf Hagedorn, Leon Van Hove, T. D. Lee, André Martin, Louis Michel, Walter Thirring, C. N. Yang, Bruno Zumino.

Still, CERN at that time was mainly in the business of strong interactions and given my most recent work on the quark model it was natural that my preoccupations focused on this topic. The person directly involved in the quark model was Van Hove (in my previous work, as well in some work done a short time later, during my stay in the US, I used some of his results). However, his, as well as my interests started to shift, so I don't remember to have had, during my first CERN stay, many discussions with him, mainly due to the fact that Van Hove was not an easy discussion partner, despite, or perhaps because of his remarkable intelligence. (Many of my colleagues who knew him had the same impression). But our preoccupations continued to partially coincide for many years, and our convictions, too. Van Hove was a firm supporter of the idea that Quantum Chromodynamics (QCD) could not replace the phenomenology of strong interactions, because quarks and gluons are under normal conditions confined and not accessible to direct experimentation. However, when I started in the early 1970s to apply statistical methods, Van Hove, although a renowned expert in statistics, was still skeptical about the applicability of statistical concepts to particle physics. Eventually, however, he also joined the statistical camp, a choice his CERN colleague, Hagedorn, had made long time before, and participated actively in the quark-gluon plasma venture. Because of these common interests, whenever we met at CERN or at conferences, we had an exchange of opinions on physics problems, but also on more general subjects.

Strong Interaction Phenomenology; Regge Poles and Duality

The properties of elementary particles manifest themselves in the way they interact. This interaction can lead to bound states that is stable systems like atoms or nuclei, which we find in nature. In the accelerator laboratory, however, we also study another manifestation of the interaction, the scattering of one particle by another, similar to the scattering of one billiard ball by another. A two particle scattering process is characterized by the total energy of the colliding particles and the momentum transfer, from the projectile to the target. A full-fledged theory must be able to predict not only the probability that such a scattering process takes place, but also the way this probability depends on these two variables. (While classical physics deals with certitudes quantum physics deals with probabilities only. The magnitude of the scattering probability, usually expressed in terms of the cross section, depends on the strength of the interaction: the stronger the interaction the bigger the cross section is). While for electromagnetic interactions such a theory, QED, has been available since the 1950s, for strong interactions this is not the case even today, because QCD suffers from the confinement pathology mentioned above. That is why for strong interactions one uses phenomenological *models*, which are based on analogies and which describe only certain aspects of these processes. One of these models was the *naïve* quark model, which reduced the scattering problem of two hadrons to that of its constituents, the quarks. In the late sixties this model seemed to have reached its limits.

For experimental reasons — too low energies for multiparticle production — in the 1960s particle physics, with the exception of cosmic rays physics, was centered on two-body reactions. In analogy to quantum electrodynamics in which scattering processes between two electrically charged particles take place through the exchange of photons, in strong interactions one assumed a similar mechanism, the role of photons being taken over initially by the pion meson, the only hadron known besides nucleons. With the discovery of other mesons, called resonances, because they were more massive than the pion and thus appeared as higher frequencies of a vibrating string, this picture had to be modified. Rather than considering separately all the possible meson exchanges a short cut was invented, called Regge pole, which was supposed to describe mathematically the interaction process. The Italian physicist Regge had found that the scattering probability — actually the scattering amplitude, which is proportional to the square root of the probability — had certain mathematical properties called poles, which summarized the exchange of several resonances. Alternatively, instead of considering the exchange process, the scattering could be related to the direct formation of resonances. This equivalence is exploited by the dual resonance model. With some exceptions, strong interaction theory at CERN during that time was dominated by Regge phenomenology and the duality concept.

I had not been an enthusiast of Reggeology because it seemed to me too strongly based on analyticity postulates and also because it predicts only the energy dependence of the scattering amplitude, leaving the dependence on momentum transfer arbitrary. Furthermore it referred only to two-body scattering process and did not address, what I considered the main task of strong interaction physics, the many-body problem. Indeed the main characteristic of strong interactions is the multiparticle production process: while in electromagnetic and weak interactions the mean number of newly produced particles is always much smaller than unity, in strong interactions, this number can be arbitrarily large. Actually, I expressed my doubts about what was in the late 1960's and early 1970's *mainstream* high-energy physics at the 1972 Batavia Conference. At that time I was rather an exception, although in good company: Heisenberg, Fermi, Landau and Hagedorn, when dealing with strong interactions, considered mainly many particle production. However, as far as I remember, at that conference or very soon after Peter Carruthers expressed his support for my point of view (cf. Chapter 12) This was the start of the collaboration and friendship between the two of us, which lasted until his untimely death.

From a historical point of view it is worthwhile mentioning that at present one can hardly find a particle physicist below forty who knows what Regge physics is about, since two-body hadron reactions do not make part, anymore, of mainstream physics and also because hadron models have been substituted by QCD-inspired phenomenology. Still, Regge phenomenology and the dual-resonance model played an important historical role, among other things because they lead, through the work of Veneziano and collaborators, to the development of string theory, which at present some physicists consider the most promising candidate for the "ultimate" theory of particle physics, the theory of *everything*. We witness here what I am tempted to call the Münchhausen principle and which will be encountered also in other sections of this book.

The Münchhausen Principle

In one of his adventures the famous Baron von Münchhausen climbed to the sky on a ladder. Once he reached the top of the ladder he lifted it and started the game again. When he finally arrived at the destination, the preceding steps, without which he would never have succeeded, did not matter. Applied to science the Münchhausen principle states that *what matters is where you arrive and not where you start*. Indeed, in the history of physics there have been many phenomenological models, which served as initial motivations or intermediate steps and which eventually were forgotten, being replaced by the results they had led to and which then were raised to the rang of new postulates or models. To quote what is perhaps the most famous example in this sense: Maxwell thought that light propagated through a luminiferous aether with very strange if not contradictory properties. This did not prevent him in deriving his famous equations, which unified electricity and magnetism, which reduced light-waves to electromagnetic ones and which constitute at present *the* theory of electromagnetic interactions.

10

Statistical Concepts in High-Energy Physics; Phase Transitions

---

Despite the dominance of Reggeology and duality there was at CERN, however, even at that time, in the late sixties, an exception to this trend. This exception was represented by Hagedorn's approach to strong interactions, based on his statistical bootstrap model. Statistical methods had been used in this field also in the past, among others by Bohr, Bethe, Heisenberg, Weisskopf in the thirties and Koppe, Fermi, Pomeranchuk and Landau, in the fifties, but later on went temporarily out of fashion, in part because of other phenomenological developments which seemed more promising (these expectations were rarely confirmed) and in part because theoretical physicists were interested in theories of more fundamental nature.

Statistical physics deals with probabilities and uses, like its classical cousin, thermodynamics, notions like temperature and energy. Actually the concept of temperature permeates indirectly high-energy phenomenology through the notion of fireball. A fireball is an analog of a macroscopic "burning" system and represents in reaction theory the intermediate system, which emerges immediately after the collision.

Hagedorn's bootstrap theory can be summarized by the sentence: "A fireball consists of fireballs, which consist of fireballs, which consist of..." This theory fascinated me because it seemed to me the only approach in which the very nature of strong interactions, multiparticle production, was taken into account and effectively exploited. Although I myself never applied the bootstrap model, Hagedorn's approach influenced my later work in the sense that it convinced me that strong interaction phenomenology had to use statistical methods. As a matter of fact, my only incursion into the Hagedorn bootstrap happened much later, when Gerald Fowler and I proved[43] that the Hagedorn *exponential* mass spectrum was incompatible with the existence of quark matter, as it was conceived at that time, namely a weakly interacting, ideal gas-like state of quarks and gluons. Conversely our proof implied that the existence of such a quark matter phase meant that the hadronic mass spectrum could not grow faster than power-like. This put under question mark not only the bootstrap model but also the dual resonance model, which also predicted an exponential mass spectrum.

Statistical methods were introduced in physics after the discovery of the atomic structure of matter. The huge number of atoms, typically of the order of 10^{23}, implying an equivalently large number of degrees of freedom, which cannot be treated individually, made this approach to *macroscopic* systems absolutely necessary. It explained and replaced in great part thermodynamics, providing a statistical interpretation of thermodynamical concepts like temperature, pressure, equation of state and equilibrium. The applicability of statistical methods to nuclear and particle physics is less obvious, given the huge difference in the number of degrees of freedom. This was a major heuristic difficulty, which statistical approaches to particle physics had to face. Bohr, Bethe, and Weisskopf used in the 1930s the statistical approach to strong interactions involving nuclei assuming that the number of degrees of freedom, assimilated with the "level density," was sufficiently large. Later on Heisenberg, Koppe, Fermi and Landau went one step further and applied the statistical approach also to those particle reactions in which the number of *produced* particles was large. The implicit assumption behind these applications is that in physics anything bigger than 3 means large. This assumption has in many circumstances proved to be fruitful. Hagedorn went even further, by applying statistics also to two-body strong particle reactions, including elastic scattering. He did this by arguing that in strong interactions the level density is represented by the mass spectrum of hadrons, which strongly increases with the mass. As a matter of fact, with the advent of QCD and the discovery of the quark and gluonic structure we have learned that the number of degrees of freedom in strong interactions is much larger than initially thought, even without taking into account the mass spectrum of hadrons. Last but not least, in the 1970s with the advent of the standard model statistical methods were applied with success also to weak interactions, leading to new insight into the evolution of the early universe following the Big Bang (cf. chapter 13).

In strong interactions particle physics there exists another serious argument for the use of statistical concepts. It is an argument, which historically lies at the origin of the application of statistical concepts to these reactions and it has not lost its justification even now, more than half a century later. The argument is of experimental nature and refers to the observation that the transverse momentum distribution of particles produced in multiproduction high-energy particle reactions has a cut-off in the sense that high transverse momenta are very rare. The distribution has an exponential shape (Boltzmann distribution), which obviously suggests statistical equilibrium. The cut-off of transverse momenta, discovered for the first time in cosmic rays and later confirmed in accelerator experiments, has not found, up to the present, other satisfactory explanations, but in statistical terms. According to this statistical interpretation in a high-energy hadronic collision a thermally equilibrated system ("fireball") is created, which, in its own frame, decays isotropically. The effective temperature of the fireball is in a first approximation independent of the energy of the reaction and, as proven by the experimental transverse momentum spectrum, its magnitude is of the order of 140 MeV. Here from Hagedorn's idea that this is the maximum temperature of hadronic matter. (For Landau and Pomeranchuk, however, who were the first to put forward this statistical interpretation, the value of this temperature is determined by the mass of the pion, the lightest meson).

However, Hagedorn did not stop here, because from his statistical bootstrap theory follows that this is the maximum temperature of matter in general. That was quite an astonishing result, which contradicted conventional statistical and thermodynamical concepts. This contradiction was one of the reasons why I never applied the statistical bootstrap. Another reason was that I considered that Hagedorn had already made the best out of this beautiful idea. Still the maximum temperature preoccupied me and ultimately, made me look for alternatives. Actually, from the moment I read Hagedorn's papers, I suspected that his maximum temperature was rather a critical temperature, characterizing a phase transition. The question was, however, what were the phases involved. At that time nobody suspected the existence of quark matter, at least in the sense it is understood now. Soon after I left CERN and came to Indiana Ken Galloway, Ady Mann and myself found in the experimental data some evidence suggesting that hadronic matter had superfluid properties and that this superfluid underwent a phase transition at a critical temperature, which was numerically quite near to that found by Hagedorn (cf. chapter 12). This finding seemed to answer in part the question what the phases were, if Hagedorn's temperature was indeed a critical temperature.

With the emergence of QCD and the idea of deconfinement of quarks and gluons the significance of Hagedorn's temperature as well as that of superfluidity of hadronic matter have become much clearer. At present there seems to be agreement that Hagedorn's maximum temperature is in fact a critical temperature, characterizing the phase transition from hadronic to quark-gluon matter.[44] In the same context the superfluid phase has found its microscopic description in the quark and gluon condensates.

The initial separation between duality and statisticalconcepts did not last very long. Four years after the publication of Hagedorn's first paper on the bootstrap, where the exponential mass spectrum was derived, several authors showed that the dual resonance model also lead to an exponential mass spectrum and after another four years I suggested another link between duality and statistics in an approach called statistical duality (cf. appendix 1). Still, from Hagedorn's point of view, this was a very long time. During this time he felt quite isolated and, as far as I understand, even later continued to represent a "singularity" at CERN. But as often is the case with singularities, his contribution to the development of theoretical physics was very important. Indeed, with the start of the search for quark-gluon plasma statistical and thermodynamical concepts became indispensable. One of Hagedorn's merits was that he had kept alive among high-energy physicists the interest in these concepts. Without his contribution the progress in this field would have taken much longer. Actually, as far as I know, he coined the term *hadronic matter*, emphasizing the macroscopic properties of this state of matter, which played an important role also in the evolution of the universe. Once this concept was established it became much easier to accept the notion of quark matter or quark-gluon plasma, which in the evolution of the universe preceded that of hadronic matter. Whenever I came to CERN, in the seventies and eighties, before Hagedorn's official retirement, and also after that the door of his office was the first one I knocked at. I am happy to be able to say that I was always welcomed and I have been grateful for this privilege. We used to discuss first of all physics, but not only physics. I remember among other things that when he tried to explain to me his love affair with statistical methods he quoted once his Göttingen teacher Abraham who had told him that when you will get older you will start to like two things: red wine and thermodynamics.

I also met Hagedorn quite a few times at international physics meetings. On the occasion of one of these meetings (I think this happened in Karlsruhe) he invited me to dinner where I also met his wife. We had an interesting discussion about present Germany. Both Hagedorns, who lived in France near Geneva wanted to know how somebody with my background felt in Germany. Mrs Hagedorn had relatives with a similar background, but who apparently did not survive.

Section IV

Bonn, Bloomington (Indiana), London; Superfluidity of Hadronic Matter (1970–1974)

11

Bonn

Initially I was supposed to spend my Humboldt fellowship at Aachen University with Kastrup, whom I had known by correspondence and whose interests overlapped in part with mine. (I later met him in Bratislava, at the triangle meeting, short time before our escape). While at CERN I decided to go to Bonn, not so much because physics at Bonn was more related to what I have been interested in, but rather because in the mean time I met Wolfgang Paul, the future Nobel Prize winner and at that time the head of the Bonn Physics Institute who invited me to come to his institute. Paul who knew my situation had told me that due to my nuclear physics background I had good chances to find at the end of my Humboldt fellowship a permanent job at the Jülich Center for Nuclear research, one of the most important research institutes of Germany, about an hour from Bonn. This center was associated with the University of Bonn and Paul was a member of its governing board.

I arrived in Bonn at the end of February 1970. Before that, however, my Humboldt award was formally transformed into a contract with the University of Bonn, because the awards were reserved for foreign scientists who had the intention to return to their countries of origin. This, obviously, did not apply to my case, because my publicized escape from Romania had made clear that I would not return and the Humboldt Foundation was afraid of diplomatic complications, endangering its program in Eastern Europe. This resembled to what had happened at CERN.

My first impressions of Germany were of linguistic nature: I found out that the German I spoke contained a few Austrian idioms, which many people did not understand, e.g. the Austrian Paradeis instead of the German Tomate, Putzerei instead of Reinigung (cleaning) etc. Even now, after more than thirty years of continuous life in Germany, our daughter Diana keeps discovering some of these Austriacisms in my German.

Although the capital of the biggest country of Western Europe, Bonn was in 1970 a very provincial city, not so much by size (almost 300,000 habitants, including the surroundings) but by character. Indeed, Geneva, with a smaller population and just the capital of a canton, was much more international. Besides the government, the parliament (Bundestag), the university and a few cultural institutions with federal character like the Deutsche Forschungsgemeinschaft (the equivalent of the US National Science Foundation) and the Alexander von Humboldt Stiftung, there was essentially nothing I can remember. I was also surprised by the difference between the image of Germany of the sixties, seen from outside, and reality. This difference, which I experienced briefly, during my six-month stay in 1970, and in greater length during the years following my return in 1974, has preoccupied me and I will discuss it in some detail later on.

The University of Bonn reflected in a certain sense the provincial character of the city. For European standards this university is not very old (it was founded in 1777) and it could not compete, by famous names of professors and students, with other, older or even newer German universities.

In 1970 there were two physics institutes in Bonn. There was the Physikalische Institut on Nussallee 12, with a more general character, under the directorship of Paul. I belonged to the section of theoretical physics led by Horst Rollnik, but he had been sick quite a long time, so I saw him only near the end of my stay. The institute had a small electron accelerator, where high precision low energy experiments were performed. Vector dominance was the name of the game. As far as I can judge, it was serious, but, from my point of view, not very exciting physics. However, a few years later, when I started to use the sigma model I came to appreciate the importance of that work. In 1970 I continued my work on the quark model and interacted only occasionally with the other theorists.

This non-interaction applied also socially: With the exception of W. Pfeil, a theorist who had helped us settle, find a furnished flat in Oberkassel, a suburb of Bonn, buy our first car, and who had invited us also to his home, and an Indian theorist from Madras, with whom I shared the office, I do not remember having had any social contacts with my German colleagues of the Physikalische Institut, perhaps also because of the shortness of our stay there. Fortunately we found an old friend and colleague of mine from Bucharest in Bonn, George Comsa and his wife Gerda, who happened to be at the Institute of Physical Chemistry, also with a Humboldt fellowship, and we spent with them much of our free time. Eventually the Comsa's also decided not to return to Romania and joined the Jülich center.

Quite soon after our arrival in Bonn, Uli found a job as a chemist in a dye factory in Siegburg, a suburb of Bonn. One of the few things I remember of Bonn quite well is the car, among other things because it confirmed what our colleagues at CERN had told us, when they learned that we want to buy a used car in Geneva: used cars you buy in Germany. Indeed the car we bought was a very old beetle, for which we paid 500 DM, a ridiculous price, even at that time, and which we sold, after six months, when we left for the US, for the same amount. We never had problems with it and we still have some nostalgic feelings over it, because it permitted us to visit the beautiful surroundings of Bonn and the wonderful, fairy-tale Rhine valley and the action place of Heine's famous Loreley poem. Of course, seeing this place reminded us how the Nazis treated the work of this greatest lyric poet of German literature.

The other physics institute in Bonn was the Institut für Theoretische Kernphysik, located next to the Physikalische Institut, on Nussallee 14. Its director was Bleuler. Besides Paul, whom I had met at CERN, Bleuler was the only Bonn physicist whom I had known before coming to Bonn, and that only by name. Soon after my arrival I contacted him. To my pleasant surprise he invited me immediately to his house, where I was impressed to see hoisted on its front the Swiss flag. When I asked him how he felt, as a foreigner in Bonn, he said: "Bonn is a wonderful place, there are five trains a day to Switzerland." This remark and the discussion, which followed confirmed certain apprehensions that I had started to have about Germany after my arrival. I slowly realized that the image of Germany I had had until then, had to be qualified.

During my Humboldt stay in Germany I paid a visit to Heisenberg, in Munich. Heisenberg was at that time the president of the Humboldt foundation (the Humboldt award I had received in Bucharest was signed by him), but my meeting with him was more than a formal visit, because in some sense I saw in him my scientific grandfather. As mentioned before, my PhD adviser, Ţiţeica, had been one of the first PhD students of Heisenberg in Leipzig. As far as I could judge, in 1970 Heisenberg wasn't scientifically active anymore, but he was familiar with what was going on. We discussed the quark model, since this was the topic I was working on at that time. One of the big puzzles preoccupying high-energy physicists was the fact that quarks had not yet been experimentally seen. For Heisenberg, however, this was no surprise. He, like Gell-Mann and Zweig, considered them rather as pure mathematical entities. The postulate of confinement had not yet been formulated; thus the less the idea of deconfinement at high temperatures.

My first, indirect, impressions of post-war Germany dated essentially from 1965, when I met some Germans in Romania at Eforie on the Black Sea, who were spending their vacation there. Among others I met and befriended an elderly couple, Kurt and Emmy Schöpflin. The closeness of this friendship, which lasted until Kurt's death and which has now been extended to his daughter Gisela and her husband Klaus Hertwig, can be judged from the fact that one day before passing away Kurt asked his family to call me from Marburg to the hospital in Hannover, where he was expecting his death, to say good-bye.

Kurt Schöpflin's father had been the first social-democrat deputy in the German Reichstag. In his parents' house he had met all the important political leaders of that time, in particular the social democrats of pre-Hitler Germany. His "connections" would play a role in my settling in the Federal Republic. Kurt himself, a journalist, had been persona non grata during the Nazi period. He had worked as a simple setter in a printing-house and, being "unreliable," had not been taken to the army. Actually, he considered those (few) who, like Willy Brandt, had emigrated because of political reasons, almost as traitors. This negative judgment of his did not apply to Jews, because they had to run to save their lives. A few years later, when I established myself in Germany, I discovered that this point of view was shared by many people of his generation. Until 1990 the city of Berlin had refused to name a street in honor of Marlene Dietrich, who had turned her back to Nazi-Germany, and many Berliners opposed when their city decided to bury her in her hometown Berlin.

This attitude of the senior, post-war generation has two reasons. Those very, very few who, like Kurt Schöpflin, were convinced Anti-Nazis and who had stayed on, believed that it was their duty not to leave their country in those difficult times. The vast majority, however, had a much simpler reason for their staying on: up to a certain moment they had, in many ways, benefited from the Nazi regime. Not only had the regime apparently solved the unemployment problem but it also offered them a kind of moral and political compensation for the defeat suffered in World War I, not to mention that quite a few "Arian" Germans profited directly from the expropriation of their Jewish fellow citizens. That is why many Germans considered those who had emigrated as their enemies.

Another curious aspect of German after-war mentality was the tendency, which one finds even today in certain circles, to attribute all the evils in the second half of the 20th century, including the division of Germany, to communism. These people conveniently overlook that had a Nazi regime in Germany not existed and had this regime not started the Second World War, the Soviet empire would have never been able to extend itself up to the Elbe River. A related phenomenon has been the tendency, among the older generation and even among some historians, to compare (often in numbers of victims) the crimes of the Nazi regime with those of the Communist regimes. To many of these "accountants" this comparison offered consolation and even an excuse for the crimes of their parents or grandparents. Not to mention revisionist "historians" who denied the scale and even the very existence of Nazi extermination camps.

Interestingly enough, in the fifties and sixties outside Germany these political shadows were not easily visible except, probably, for experts. In Romania, besides listening regularly to the *BBC* and *Radio Free Europe*, I started to listen in the late sixties to *Deutsche Welle*, the official German World Broadcasting Service. I also used to read *Le Monde*, that one could buy in the late sixties in certain hotels and the German weekly *Der Spiegel*, which I got thanks to Kurt Schöpflin, who had arranged for me a subscription. My overall impression had been that the Federal Republic of Germany was a democratic country and that its population had put a definitive period sign after the chapter 1933–1945. Therefore I was rather surprised when I found out in the 1970s and even later, by direct contact with everyday German life that, although in very rough lines this impression had been correct, the detailed pattern was much more complex.

The meeting with Bleuler, who lived in Germany for some years already, contributed to the completion of my picture of Germany, because it was not only the provincial atmosphere that made him complain about Bonn, but the fact that the Nazi past still pervaded its present. In essence we both agreed that many, if not most of the Germans had not yet really understood what the period 1933–1945 had represented for them and for the rest of the world. Even more astonishing was perhaps the fact that many of the Germans I discussed this issue with, and the democratic convictions of whom I had no reasons to doubt if for no other reason but because of their age, did not realize how different their perception of the past, in particular of their past, was from that of other nations. As mentioned, the age played a role in this perspective since the majority of the older generation, which had been directly involved in the war, tried desperately to repress this chapter of their life. This repression led to different strange effects, (cf. chapter 16), among them the "we didn't know" syndrome, as if the *Kristallnacht* and the following deportations of Jews would have happened on another planet. And if you come to think of it, in my first week in Bonn I found in the middle of the city under a bridge a plaque reminding that in that place there had been the Jewish cemetery, which had been destroyed by the Nazis during their rule.

Actually, this happened with most of the cemeteries in Germany (Marburg is an exception) reflecting the will of the regime not only to physically eliminate the Jews, but also to eliminate their remembrance. While the physical extermination succeeded almost completely, the spiritual was less successful. Heinrich Heine could not be eliminated from the German literature, nor could Mendelsohn from the German music — without Mendelsohn even Bach might not have been known today — not to mention the contribution of Jews to German science. And even the signs of the physical presence of Jews in Germany could not be entirely eradicated. In the archeological museum of Cologne there are for example gravestones, which witness the presence of Jews on German soil before German tribes settled there — the first Jews came with the Romans.

12

USA

Sometime in the middle of my Humboldt fellowship I got two job offers from the United States. One was a visiting associate professorship from the University of Pittsburgh and the other a visiting full professorship from Indiana University. I had thus, essentially, five alternatives: to continue to stay in Bonn, where Paul had already made financial arrangements, to go to Tel Aviv, to go to Lyon, or to go to one of the two American universities. For somebody who had just escaped from behind the Iron Curtain and who during almost eleven years had to do research as a private person, and that under quite difficult circumstances, without the possibility to travel and to attend meetings, it was a real *embarras de choix*.

The American offers were very tempting, first of all because the USA was the forefront of physics and also because I wanted to know the New World. As a matter of fact I had already gotten a foretaste of this New World attending a summer school at Brandeis University in Boston. There I met among others Henry Primakoff, with whom I had common scientific interests and with whom I got immediately befriended. We spent many hours discussing physics and life and a year later, while in Philadelphia, I visited the Primakoffs at their home. At this school I also met Steve Weinberg. Another participant and lecturer I remember was Maurice Jacob, whom I knew from CERN.

On my way to Boston I stopped in New York where I visited my old friend Joe Policano, whom I had met for the first time in the sixties in the sleeper traveling from Bucharest to Budapest. In New York I also tried to contact some relatives from the side of my mother. The sister of my mother, who had emigrated to the USA before World War I, did not live anymore, but her husband and her son, as well as a cousin of my mother, lived in New York. I succeeded in meeting only that cousin, a lady in her eighties, who had left Europe at the beginning of the century. With my uncle I only talked on the telephone; he was too old and sick to meet me. I never managed to contact my first-degree cousin, the son of my aunt, Robert Engler, who was a university professor.

(*Note added in proof*: A few days ago I managed to find Rober Engler, alas too late. Working on the proofs for the present book it came to my mind to do a Google search on Robert Engler. To my surprise and deep regret I discovered three obituaries about him in *The Washington Post* (March 6, 2007), *Boston Globe* (March 11, 2007) and *The Nation* (April 2, 2007) and then through the newspapers I got in touch with his family, who was as surprised by this postumous meeting as me. Robert Engler was a well known political scientist and author of many important books, in particular about the political aspects of the oil problem.)

One detail of this journey I remember and which impressed me was the meeting, on the transatlantic flight, with an American much older than myself, who had spent a long time in Germany and who happened to be my row neighbor. When he heard that I was visiting the US for the first time and even considering taking up residence there, he told me: life in America is, from certain points of view quite similar to that in Europe; there, as in Europe, people spend more money than they can afford. What is quite different is the country itself: You still find woods there that are totally unexplored and which have never been penetrated by men, with the exception of Native Americans, perhaps. The emotion you feel when walking through such a forest was, according to my plane companion, extraordinary, unforgettable. Although I never had the occasion to go through such an experience literarily, I thought that I understood what he meant:

I referred it to the process of discovery, the discovery of new lands and continents, the discovery of the New World. It explained to me the pioneering spirit of the American people, their curiosity and drive to find and learn new things. But it also suggested to me an analogy pertinent to the present book: I associated it with the act of discovery in science, with the joy of insight, with the sensation a researcher has when he realizes that he is the first one who has found a particular new phenomenon or a particular new relationship nobody before him had seen or discovered.

I also remember quite a different detail of this journey, the trip from New York to Boston. Not only was the TWA flight very late, (I arrived in Boston around 2 a.m.,) but I was the only passenger in the Boeing 707. No wonder that TWA, eventually, went broke. Still, although 2 a.m. Levy Rahmany, my old friend from Focşani and Bucharest, with whom I had shared a bed during our student years, was waiting for me at the airport. He and his family, which I had not met before, had come to Boston from Israel for a 5-year stay in Boston to continue their professional specialization, he as a psychologist and Rodica, his wife, as a MD.

Another small detail I remember from this trip is the fact that bus stations in Boston had no distinctive sign as they have in Europe, except a red spot on a telegraph pole. There was no mentioning of the number of the bus, its schedule or any other information. Sometime later I would discover even railway stations, like that in Santa Fe, without signs, not to mention that these stations had no buildings, ticket counters or schedules. All over all I found the difference in civilizations between the old and the new continent much bigger than that between Eastern and Western Europe, despite the fact that the political and social systems on both sides of the Iron Curtain were so different, while those in Western Europe and America were quite similar.

Indiana University

Shortly before the Brandeis summer school I had decided to accept the Indiana offer, among other things because Hardy Mayer, who had been a professor there before joining the University of California at Irvineand who had introduced me to the Indiana physics department, had told me that Bloomington, Indiana was a nice place, with nice colleagues and very good music. The Music Department of Indiana University was the biggest music department in the USA. Besides a good opera — some of the department professors like Eileen Farrell were famous singers (she had been for years a Prima Donna at the Metropolitan in New York and other famous operas) — there were renowned musicians like Zino Francescatti and Janos Starker and the excellent Berkshire Quartet. Bloomington, due to its magnificent university campus also turned out in other respects to be a very pleasant place. Although we were fresh immigrants, we were accepted immediately by the American society and colleagues and neighbors treated us very friendly. Our daughter *Diana Free* was born here. We chose her name as a reminder that this had happened in Indiana and that she was a free US citizen. We, too, got permanent residence almost immediately because, being born in Czernowitz, that belonged in 1970 to the USSR, I benefited from the unused immigration quota.

The Physics Department in Bloomington was quite big — it had about 40 professors — and essentially all fields were represented. Among the theoretical physicists I interacted with besides Ady Mann, with whom I actively collaborated, were Roger Newton, Don Lichtenberg and Barry Malik. I also met and became friends with Josef Jauch, who spent a few months as a visitor in Bloomington. Two years later in Vienna I met his collaborator, Fritz Rohrlich with whom he had published a classical book on field theory. Another remarkable detail I remember was the common tea with the Mathematical Department; here I met Max Zorn, the author of the famous Zorn lemma, who was already retired and Andrew Lenard. As a matter of fact, among the things I would later miss in Marburg was also the tea institution so characteristic for Anglo-Saxon countries. After Bloomington, I found it of course again at Imperial College in London.

During the period 1970–1972 I visited many physics departments and laboratories in the USA. Among others I spent three weeks at Brookhaven National Laboratory, at the invitation of Maurice Goldhaber, the director of the laboratory. Maurice, with whom I befriended, had worked on isomers and he introduced me also to his colleagues who were familiar with my work on isomeric shift and beta decay. I also visited the West coast, among others Hardy Mayer at Irvine. I met Roy Glauber once or twice at Harvard and Vicki Weisskopf at MIT. In 1972 I participated in the Batavia conference on High Energy Physics. This was the first major conference of my career, another confirmation that my 11 years persona non grata status in Romania had ended. I remember two details about this meeting, a scientific one and a social one. During one of the sessions I had expressed my skepticism about the dominance of Regge phenomenology in strong interactions and emphasized the importance of statistical approaches. In his concluding talk Murray Gell-Mann disagreed with my point of view, but there were others, like Peter Carruthers, who agreed. Only two years later Carruthers published his paper on "Heretical models of particle production"[45] in which he presented evidence that statistical and hydrodynamical models described the data quite well. But even before that the seminal paper by Kirzhnits and Linde[46] on phase transitions in the early universe (cf. below) shifted the attention of the high-energy community towards statistical methods, which, combined with the field theory of strong interactions (QCD), ultimately lead to a new chapter of high-energy physics, quark matter. Still, given the possible implications of Regge physics via string theory Gell-Mann may have been right, too.

During the Batavia conference an extremely grave terrorist act took place: the killing by Arab terrorists, at the Summer Olympics in Munich, of 11 members of the athletic team of Israel and of one German policeman. The entire conference was under shock; I learned about the event before going for lunch to the cafeteria and when I arrived there I looked for somebody to share the news. The first known people I saw were the German delegation, among them Paul and Rollnik, and I joined them. I had not seen them since I left Bonn in 1970, so that our encounter was quite warm. After discussing the terrible event we exchanged personal news, among other things the fact that I had become a father. During lunch I realized that although there were still many vacant seats at our long table, nobody joined us. I had the distinct impression that people avoided the German table which may have contributed to the warmth of the reception of me, who had not participated in this apparent boycott. Anyway, I had not intended to attribute any special significance to my gesture since I was sure that my German colleagues were as outraged as anybody else by that event. And perhaps even more than anybody else, because it had happened in Germany.

From that first period of my US stay I also remember my participation in the 1971 Summer Institute in Boulder, Colorado, which gave us the possibility to see a great part of the American continent, since we went from Bloomington to Boulder by car, with our three month-old baby. The beauty of Colorado impressed us very much, although the Rocky Mountains were so much different from the majestic and more savage Alps or the Carpathian Mountains, which we knew from Europe. The fact that one could drive up to the top of the Rockies at an altitude of 4000 m without finding snow, which in the Alps covers permananetly the peaks, reminded us that Colorado and the entire USA are situated much more to the South than Europe. At this institute, at which Ady Mann also participated, I met Geoffrey Sewell, who had come to lecture from London. During our London stay in 1972–1973 we also met his wife Robyn and became friends, a friendship which has continued until now. The Sewells visited us in Marburg and we visited them every time we came to London. Geoffrey also participated in the first LESIP workshop dedicated to local equilibrium in strong interactions, which I organized in Bad Honnef (cf. chapter 17).

Although my family and I enjoyed our two years stay in the United States very much, we had to think about the future. In 1972 I was 42, which is a Methushael age for America, and had no permanent job. My visiting appointment at Indiana University, initially for one year, had already been extended for another year, which was quite an exception, and the situation in theoretical high-energy physics in the States was at that time changing from bleak to black. The crisis had already started in 1969 and slowly affected the entire Western world. My job problem was aggravated by the fact that I could not conceive doing something different in physics (not to mention in another profession) than theoretical particle physics.

A Letter from the White House

My colleagues in Indiana, and in particular Emil Konopinski tried to help. Emil, a prominent nuclear physicist who had participated during the war in the Manhattan project, told me that shortly after the war he had been in a similar situation, albeit at a younger age, and that he, eventually found a job by sheer accident. He advised me to do what he had done, namely to write dozens and dozens letters of application. However Emil did not limit himself to this. He suggested we both write separate letters to Richard Nixon, at that time the President of the US, describe to him the situation and ask him for help. A few weeks later I got from Edward David Jr., the Science adviser of the President and Head of the White House Office of Science and Technology, a letter acknowledging that the president had asked him "for action."

This "action" did not materialize, except that I got a letter from the Princeton Institute for Advanced Study suggesting that I apply for a visiting position there for the year 1973–1974 following my London stay and this application did not materialize either. Actually, I did not wait too long to hear from the President's advisor since the preliminaries of the Watergate scandal began to become known and I became convinced that the US government had more urgent problems to solve. As a matter of fact, one of the first changes within the US administration affected the Office of Science and Technology. In December 1972 David resigned and the Office was subsequently abolished and its functions transferred to the National Science Foundation. In March 1973, while already in London I got a letter from the Executive Office of the President signed by Frank R. Pagnotta informing me about these developments and concluding: "I am sorry we are unable to be more helpful to you, but I am sure that you can appreciate the situation."

Fortunately I had been in the lucky situation to have the possibility to return to Western Europe and to continue to work there as a theoretical physicist. First of all Paul had offered me a job in 1970 and had reiterated his offer in 1972, when we had met at the Batavia High Energy Conference. Secondly, I had an invitation to join for one year, as a senior visiting fellow, Imperial College in London. Last but not least, I still felt more European than American and Europe attracted me very much both from the social and cultural point of view. Furthermore it was quite questionable whether Uli would find a job in the States in the same place as I. Actually in the first year of our Bloomington stay she almost found a job as a chemical engineer with Lilly, a big pharmaceutical company in Indianapolis, which was less than an hour drive from Bloomington. However in the last minute her appointment did not go through because in the building she was supposed to work there were no toilets for women: a female chemical engineer was something quite unusual in the US. A year later she found a job in Chicago with a dye company, where this toilet problem did not exist, but I had no job in Chicago... We knew that in Europe the job situation for women in certain professions like engineering was much better. For all these reasons, despite the fact that quite a few senior and influential American colleagues like Sid Drell, Maurice Goldhaber and Vicki Weisskopf encouraged me to continue to look for a job in the States, in summer 1972 I accepted the invitation to Imperial College.

THE WHITE HOUSE
WASHINGTON

August 24, 1972

Dear Professor Weiner:

Your letter of June 8 to President Nixon, along with Professor Konopinski's, was referred to me for action.

At the moment, I have nothing definite to report except that I have taken some extraordinary measures which are now in the process of development. I hope to have a report for you in the not too distant future.

It goes without saying that I much sympathize with your situation and am quite hopeful that we can be of assistance to you.

Yours very cordially,

Edward E. David Jr.

Edward E. David, Jr.
Science Adviser

Professor Richard M. Weiner
Department of Physics
Indiana University
Bloomington, Indiana 47401

cc: Professor Emil Konopinski

The Mesonic Cloud of the Nucleon and Superfluidity

My scientific activity in Bloomington was devoted mostly to the development of a model on strong interactions based on an analogy with superfluidity. Almost immediately after my arrival there I began to interact with Ady Mann, who was at that time a postdoc from Israel and who happened to share my doubts about the Hagedorn maximum temperature. We started from the concept of mesonic — more concretely pionic — cloud of a nucleon, which existed in the literature. I personally was attracted by this concept because it was linked with an old idea dear to me, the quest for finite size effects, which had been an important factor in my incursion into the isomeric shift. While in the isomeric and isotopic shift the finite size refers to the nucleus, the electron scattering experiments by Hofstadter and collaborators, which were performed more or less at the same time as my work on the isomeric shift, had shown beyond doubt that the nucleon, too, was not an indivisible object, that it had a structure. The question was what this structure consisted of. The constituent quark model was an answer to this question and this explains, probably, why I spent in the 1960s some time working on the quark model. However this picture was not useful in the problem of phase transitions — the idea of quark matter based on quantum chromodynamics and asymptotitc freedom had not yet been formulated — so Ady and I were tempted to take another look at this problem. The mesonic cloud was a natural candidate for this attempt.

From the many-body point of view the mesonic cloud was a system of bosons. The only other boson system known at that time, He-4, had the remarkable property of superfluidity: while normal liquids are viscous and therefore meet resistance when flowing through pipes, superfluids do not, they do not suffer friction. Therefore we asked ourselves whether, by analogy, this was not true also for the pionic cloud in the nucleon. The direct test of this assumption would have implied a "measurement" of viscosity, as performed in helium. This was, obviously, out of question, because of the elusive nature of our system. However, there existed other properties characteristic for superfluidity, which perhaps could be looked for. According to Landau's theory of superfluidity, which also constitutes today the basis for the understanding of this phenomenon, superfluidity is characterized by three main properties:

(1) The existence of two types of motion, a "normal" one, with the same properties as the motion of a normal viscous fluid, and a superfluid one. The ratio of the densities of the superfluid and normal "fluids," the order parameter, depends on temperature.

(2) The existence of a critical temperature. Above this temperature the entire system is "normal" and there is no superfluidity. At this critical temperature a phase transition takes place.

(3) The existence of a characteristic excitation spectrum, which at low temperatures is phonon-like.[47] This spectrum manifests itself through a critical, non-vanishing velocity for an object, which probes the superfluid system. If the velocity of that object is smaller than the critical velocity, it will interact only with the normal part of the system and the interaction will be weaker than in the case when the velocity exceeds the critical value and it interacts with the entire system. Quantitatively, the critical velocity is given by the minimum of the ratio between energy and momentum of an excitation in the "fluid." It was this last property, which seemed to us accessible to an empirical test.

In helium the critical velocity was measured by scattering neutrons on the system. By analogy, we considered[48] scattering experiments on nucleons, more precisely on the mesons bound in the nucleon, the so-called mesonic cloud. This is analogous to the treatment of scattering on nuclei as scattering on the nucleons of which the nucleus is made of. Such an approach is valid if the incoming energy and the momentum transfer are large compared with the binding energy. As a matter of fact, scattering on mesons on the nucleon has been a widely used method for the determination of the pion-pion cross sections, within the so-called Chew-Low approach. With this analogy in mind, we went to the high-energy experimentalists in Bloomington and asked for help. It happened that Ken Galloway had some bubble chamber data, which seemed to suit our purpose. In particular we considered single pion production processes induced by various particles and attempted to interpret them as knockout processes in nuclei. For reactions where the incoming energy was much larger than the binding energy we assumed that the mesons in the cloud behaved as free particles, except that they were not on mass shell, i.e. that they had an effective, not yet determined, mass, which differed from that of a free particle. Their momentum distribution in the nucleon was taken, again in analogy to certain approaches in nuclear physics, to be statistical, except that, being bosons rather than fermions, the distribution had to be of Bose-Einstein rather than Fermi-Dirac form (cf. also chapter 22), with an effective, non-vanishing temperature. In this approach the momentum of the mesons in the nucleon determined the momentum transfer of the reaction. We then calculated the cross section for the single production process as a function of the momentum transfer and compared with experimental data, for various single pion production processes, at various incoming energies. It turned out that this simple model,[49] could explain all data simultaneously, with one effective temperature[50] and one effective pion mass; the value of this mass was within the experimental errors consistent with zero, i.e. the excitation spectrum of the pion cloud was phonon-like. The rest of the nucleon — the nucleon "core" — represented by the other degrees of freedom constituted the heat bath for this subsystem. The nucleon in its ground state, like any system in its ground state, has of course a temperature[51] $T = 0$.

These empirical findings suggested that the critical velocity was non-vanishing and thus satisfied Landau's criterion of superfluidity. As a first test of this interpretation we looked for the most obvious manifestation of the critical velocity, the existence of a velocity — and hence energy — threshold below which, as in helium, the projectile interacts only with the normal part of the system and above which it interacts with the whole system. In analogy with what happens in helium, above the threshold one would thus expect an increase of the cross section and the appearance of new excitations in the mesonic cloud. In particle physics language the most obvious candidates for these excitations would be the creation ("liberation") of a pion and of resonances. Both these processes are, indeed, associated with an increase of the cross section. A further test of these considerations included an explanation for the shift with energy of the 3/2-3/2 resonance in nuclei.

We interpreted the above findings as evidence that mesons in a nucleon, like the other boson system, He-4, have superfluid properties. We were aware that this was quite a bold extrapolation and stated it clearly in our paper,[52] where we said: "This is of course an analogy, which cannot be justified at present but by its possible consequences." To make our point even clearer we added the following quotation from a paper by Chou and Yang[53]: "It will be evident that our arguments are based on imprecise extrapolations of known physical concepts. This is contrary to a contemporary fashion in our field to pursue at great lengths extrapolations of a mathematical nature. Preference for the one approach or the other is, at our present stage of understanding of hadronic collisions, a matter of taste. While both types of extrapolations are basically no more than guesses, the fruitfulness of which will have to be decided by experimental facts, we believe that extrapolations of physical concepts have a greater chance of providing a useful orientation." I am convinced that this point of view is justified at present as well; it also illustrates another aspect of the Münchhausen effect mentioned previously.

Quite soon we could derive,[54] using Landau's theory of superfluidity and the results of our analysis of empirical data, a numerical value for the critical temperature of the order of 190 MeV, which was not very far from the value found by Hagedorn. From our point of view, this result not only confirmed our supposition that Hagedorn's temperature was a critical temperature, characterizing a phase transition, and not a maximum temperature, but it also suggested a partial answer to the question what the phases of this transition were: the phase below the critical temperature was the superfluid phase. As to the "normal" phase, above the critical temperature, it would take another four years until, in 1976, Shalom Eliezer and I understood that this was, most probably, the quark matter phase.

13

London, Imperial College

Foreign countries and their capitals had from my early childhood exercised on myself a particular attraction. This was probably because, together with my parents, even before World War II I considered myself a displaced person and did not know where I belonged. But London was on the top of my priority list for a quite particular reason: in my mind it was associated with the BBC and the resistance against the Nazi barbarity. The invitation to Imperial College was for me an opportunity to combine the useful with the pleasant. A small detail might here be worthwhile mentioning. Although we had permanent residenceship in the USA traveling abroad was a big problem. Our Romanian passports were not valid anymore because we had not returned to Romania after the expiration of our exit visa. Technically we were stateless persons and for this category of people the USA (like other Western countries) issued travel documents: In the States such a document was called *Permit to Reenter*, which as the name suggests, gave the owner the possibility to reenter the US and also to apply for entry visa in the country you wanted to visit. This last detail was a big nuisance since you needed a separate visa for each country and this visa was valid for one entry only. Furthermore the formalities associated with the visa application took quite a long time. Last but not least these visas were usually issued for short periods and did not entitle to take up a job in the country visited. For our daughter Diana Free life was much simpler: born in the USA she got immediately US citizenship and hence a US passport. Although my British hosts did everything to accelerate the formalities, I arrived in London one month later than scheduled. After arriving in London Mathews, the Head of the Physics Department of Imperial College, told me that he had to make a special case with the British immigration office, which apparently had not handled such a situation before.

During the month that I was waiting for my travel documents I was officially unemployed and Don Lichtenberg, my Bloomington colleague advised me to ask for the unemployment benefits I was entitled to. I did not do that because I considered this as a sign of ingratitude towards the country that had accepted to offer me asylum.

Among the people at Imperial College I interacted with during my stay there I should mention Tom Kibble, the leader of the high-energy theoretical physics section and Shalom Eliezer, who was at that time a postdoc and with whom I started a collaboration which would continue for several years in many different places. After London we met and worked together at the Physics department of the University of Massachusetts in Amherst, at Los Alamos and eventually in Marburg. The unified electroweak interaction, to which Tom Kibble and Peter Higgs had brought essential contributions through the Higgs-Kibble mechanism of mass generation, was not yet recognized as part of the standard model. Nor had Tom done his important work on cosmology yet. But he was already quite a busy person among other things due to his teaching and administrative duties. Still he did not mind when I frequently invaded his office with a burning physics problem and was always willing to listen and if possible to help. On one of these occasions I met in his office Joseph Rotblat, the later Sir Joseph and Peace Nobel Prize winner. With Tom I regularly had lunch in the staff cafeteria and one day I remember that he introduced me to a gentleman who, judging from his informal look, I would never have guessed to be the famous Patrick Maynard Stuart Blackett, Baron

Blackett, who had already gotten in 1948 the Nobel Prize for his fundamental discoveries in particle physics. Although retired — he died one year later at the age of 77 — he made the impression of a quite alert and well informed person. With Tom and his family we also met socially in their house and in our place. Another prominent colleague at Imperial college was of course Abdus Salam, the star of the theory section of the physics department; he would get the Nobel Prize in 1979 (together with Seldon Glashow and Steve Weinberg) for the unification of electromagnetic and weak interactions. In the period 1972–1973 Abdus Salam was rarely in London, spending most of the time either in Trieste, where he had founded the International Center for Theoretical Physics, which carries now his name, or on travel. Still, I had several interesting discussions with him among other things on the effect of temperature on particle interactions.

I used to discuss with Salam also more general topics like philosophical and political ones, but also some of more practical nature, like renting a place to live in London. The first month we lived in a hotel near Victoria Station and then we found an apartment, actually half of a house, in South West London, near Wandsworth Common, one of the nice London parks. From here I had a direct bus to Imperial College in Kensington. For a huge city like London this was quite a convenient solution, which, fortunately, we could afford, because my fellowship was quite generous. In our search for a flat we met Jack Gold by accident, who was one of the most prominent British film directors, who had an apartment for rent in his newly built house, but from where I would have had to commute by train and underground to Kensington.

Among the physicists I met in London I should mention also Eric Burhop. I am not sure anymore whether I had met him before in Romania or whether we had just corresponded, but we certainly had some physics problems of common interest related to beta-decay. He invited me for lunch to the University College where he was working and was quite interested in listening to my adventures in communist Romania. Burhop who had participated in the Manhattan Project had quite liberal if not leftist views, honored among other things by the Joliot-Curie medal of the World Peace Council and the Lenin Peace Prize. I also met and had useful and pleasant discussions with Elliot Leader from Birkbeck College and last but not least Geoffrey Sewell, from the College of William and Mary and whom I knew from the States. These three colleges as well as Imperial College belonged to the University of London.

From London I made several trips to British universities as well as to some on the continent. Kemmer, the Head of the Physics Department of the University of Edinburgh, who was an old friend of Jusziu Ausländer, invited me for a talk at Edinburgh. Here I met Peter Higgs who graciously awaited me at the airport and who gave me a tour through some famous pubs where I could taste the "true" Scottish whisky. I gave a talk at the University of Glasgow and also visited the Polytechnic school in Aberdeen. They had an opening and invited me for an interview. It was from the beginning clear to me, and perhaps also for them, that this would not lead to anything, since polytechnic schools in Britain, unlike in other countries, are pure teaching institutions. I applied for the job out of simple curiosity to see what is going on in such a school and they probably invited me to have a look at somebody with my curriculum. Given the open invitation I continued to have in Germany I would certainly not have accepted this job, had they offered it to me. I remember from Aberdeen, a nice medieval city, also the strong winds that made walking almost impossible.

I was invited to the University of Oxford by Richard Dalitz and given there the High Table honor in the big hall where all students ate. In Oxford I met again Ady Mann who had also moved from Indiana to England. I also met here Rudolf Peierls with whom Ady collaborated on topics of nuclear physics. Last but not least Gerald Fowler invited me for a colloquium to Exeter: we immediately found a lot of topics of common interest, which led eventually to a long-standing collaboration and friendship.

Trips on the Continent

Wolfgang Paul, who was at that time the acting director of DESY, the German electron synchrotron laboratory in Hamburg, invited me to give a talk there and with this occasion I discussed with him and the theorists of DESY the prospects of finding a permanent job in high-energy physics in Germany. The job market in Europe in this particular field was already influenced by the crisis in the States and permanent jobs were quite scarce. Still, my colleagues in Hamburg promised to help me and this was important, because DESY was quite influential in the process of occupying the few positions available. Actually, quite soon this promise materialized when at the University of Marburg a professorship "for life" was advertized. From Hamburg I continued by train and boat my trip to Scandinavia: During this trip I visited Denmark, Norway and Sweden. In Copenhagen I met at the Nordita Institute Saul Barshay, whom I knew by correspondence. My interest in nuclear physics was at that moment quite limited and therefore I did not contact the nuclear physicists of the Niels Bohr Institute.[a]

My first stop in Sweden was Lund, in Sweden, a short ferry trip from Copenhagen and I remember that my host Bengt Svensson waited for me at the ferry. I gave a talk at the Physics Department and proceeded then by train to Stockholm, which was quite a long journey. I had deliberately chosen the train as method of transportation to learn how the Swedish landscape looked like: I remember a never-ending succession of needle forests. In Stockholm I was the invitee of Bengt Nagel, at the Royal Institute of Technology, where I gave a colloquium. I leant that part of Bengt Nagel's family came from Poland and some of its members had had a similar fate as mine.

[a] I would meet Ben Mottelson twenty-five years later in Trento, with the occasion of a workshop on Bose-Einstein Correlations. Ben was at that time the director of the ECT, the European Center for Theoretical Studies in Trento and we met at a dinner in David Brink's house. I had used the Bohr-Mottelson model of nuclei in the fifties, but we discovered that we had also other common interests and many common acquaintances.

In Stockholm I stayed at a very fancy guesthouse, which belonged to the Nobel Foundation and which impressed me, among other things, by its precious original paintings and its antique furniture. Being for the first time in Scandinavia, I was surprised by the high prices in the restaurants, as compared with the rest of Western Europe. From my colleagues I learnt that this applied also to many other items and was due to the high taxes through which the governments subsidized the social welfare of these countries, the level of which was among the highest in the world. From Sweden I went by plane to Norway. After spending a day in Oslo where I visited the Munch museum and the impressive Vigeland Sculpture Park I went to Bergen to give a talk at the Physics Department of the Unversity. I was here the invitee of Egil Lillestol, whom I knew from CERN. (I would visit Bergen again some twenty years later with the occasion of a heavy ion reactions meeting). In the plane, which brought me back to London I met another CERN colleague, John Bell, who was coming from a meeting in Trondheim and was on his way to Belfast where his mother lived. With this occasion I got first-hand information about the tense situation in Northern Ireland, about which John was very much concerned.

In 1973 we spent six weeks in Israel at the invitation of the Technion, the Israel Institute of Technology in Haifa, where Ady Mann had been appointed, after his stay in Oxford. It was my first visit to Israel where part of my family, who had survived the Transnistria camps and the Siberia deportation lived. Some of them I had not seen for more than thirty years. Although I had read a lot about Israel, I was surprised by many things; among others the apparently "normal" coexistence of Arabs and Jews. The only *juridical* difference between these populations was that at that time Arabs were not accepted in the Israeli Army. Another surprise was the oriental impression the country made on me when driving for the first time from the airport to Tel Aviv and Haifa. I had expected a much more occidental atmosphere. In the meantime this situation has changed very much.

I liked the atmosphere of the physics department at the Technion and the people I met there. Besides interacting with Ady, I met among others Charles Cooper, Arnon Dar, Asher Peres, Michael Revzen, Nathan Rosen, who was at that time already retired and Paul Singer. I participated with pleasure in the bridge games during lunchtime. This pass time is popular among physicists and I had the occasion to practice it also at the bridge club of CERN. Actually, I remember that many of my Polish colleagues who commuted between Moscow and Dubna used to play bridge in the bus, and, last but not least, I met Uli with the occasion of a bridge party in Bucharest.

At the invitation of Walter Thirring and Herbert Pietschmann I participated in the 1973 Bratislava-Budapest-Vienna triangle meeting, held that year in Vienna. It was pleasant to visit old Vienna again, the first free city in our journey to freedom and to meet here old and not so old friends.

Superfluidity and Symmetries

During my stay in London at Imperial College I met Shalom Eliezer who was at that time a postdoc from Israel. He was working on symmetries in elementary particle physics and became interested in our results on superfluidity. Since in helium the superfluid phase transition is of second kind[55] and is therefore accompanied by a change of symmetry, Shalom and I suspected, by analogy, that this would happen also for superfluid hadronic matter. We therefore started to look for a symmetry change mechanism, which could correspond to the phase transition in question.

It was clear to us that the most natural candidate for this mechanism was that of spontaneous breaking.[56] This is stated explicitly in our 1974 paper.[57] However, at that time we were discouraged by the Nambu-Goldstone theorem, which implied the existence, for spontaneously broken symmetries, of a scalar massless particle for which there was no evidence. That the phonon, which our phenomenological considerations in Refs. 49 and 52 had brought to light, might be the missing Nambu-Goldstone boson, did, probably, not come to our mind, but two years later (cf. below). Furthermore, at that time quantum chromodynamics with its symmetries was not yet established as *the* theory of strong interactions (some of the papers, which in hindsight proved to be relevant for QCD, had barely appeared) and flavor symmetry, rather than color[58] symmetry was still considered to be the essential symmetry of strong interactions. These were the reasons why in Ref. 57 we attempted to associate the superfluid phase transition with the dynamical, explicit breaking of the SU(3) flavor symmetry in strong interactions, the more so that we knew how to handle technically this problem. We conjectured, in particular, that SU(3) was broken at low temperatures and restored at high temperatures. High temperatures meant for us large center of mass energies and high transverse momenta. The restoration of symmetry would have implied, among other things, that at high energies the differences between the properties of the Kaons and pions determined by the differences of their flavor content would vanish, which was consistent with experimental data.

Two years later, however, in a paper entitled "Phase Transitions of Second and Zero Kind; a Phenomenological Field-Theoretical Approach. Sigma model and superfluidity of hadronic matter"[59] we returned to spontaneously broken symmetries and treated supefluidity with the help of the sigma model. This model contains besides the scalar sigma particle (which plays the role of the Higgs-Kibble particle in the sense that it generates masses) pions and quarks. In this interpretation, in the superfluid phase the (chiral[60]) symmetry is spontaneously broken by the non-vanishing vacuum expectation value of the sigma field, while the vacuum expectation value of the pion field is vanishing. This last fact is reflected in the phonon-like spectrum of the pion cloud, the starting point of our superfluidity approach. Above the critical temperature the vacuum expectation value of the sigma field vanishes and therefore the quark mass (which is proportional to the vacuum expectation value of the sigma field) vanishes too.

As far as I know,[61] Ref. 59 (as well as Ref. 57 quoted above), are among the first, if not the first, papers, where the possibility of looking in *the laboratory,* in multiparticle production processes, for the changes of symmetry of strong interactions, are discussed.[62] In our 1976 paper, in particular, the link between chiral symmetry and the phenomenology of hadronic-quark matter phase transitions is studied. This phenomenology is represented among other things by the momentum transfer distribution in hadronic reactions, which can be related to the temperature of the system. In Hagedorn's original theory there is a maximum temperature and no phase transition at all. This we called phase transition of zero kind. Based on our previous work on superfluidity, we suggested, on the contrary, that at high energies there is a phase transition of second kind, where the chiral symmetry is restored. As possible evidence for this transition we quoted the *change of shape* of the momentum distribution of secondaries produced in high-energy hadronic reactions. At that time high-energy heavy ion reactions were still in the dream-phase, but we were convinced that nucleon-nucleon reactions were good enough to prove our point. As a matter of fact, even today, when heavy ion reactions have almost completely taken over this kind of physics, I continue to believe that at least certain features of hadronic reactions, like the cut-off of transverse momenta, cannot be understood but in terms of statistical concepts.

The chiral phase transition considered by us in that paper became later the central topic of quark-gluon plasma search, because this transition is considered to be either identical with, or related to, the deconfinement phase transition, and constitutes thus the portal to the new state of matter, quark matter, which is at present in the center of interest of high-energy heavy ion reactions.

A by-product of this approach to superfluidity in strong interactions was a model for the phenomenological description of confinement, developed in the eighties,[63,64] and which has become very actual in view of some surprising results of the heavy ion experiments at the Relativistic Heavy Ion Collider (RHIC): the quark-gluon plasma, which is presumably seen in these experiments is, contrary to most expectations, a strongly interacting system. This property was predicted in Ref. 64. A further indication for superfluidity in the new results obtained at RHIC is the perfect fluid behavior of the matter seen there and which manifests itself among other things in the strong elliptic flow. These observations are discussed in greater detail in chapters 27–29. Another type of evidence for superfluidity considered in Ref. 59 referred to the *scaling* properties of the structure functions in deep inelastic electron proton scattering (cf. also below).

Supefluidity and Superconductivity: Analogies and Follow-ups

In 1975 Chapline published a paper entitled "Is hadronic matter a superfluid?"[65] in which he suggested that "the scaling behavior observed in deep inelastic electron-proton scattering can be simply explained if it is assumed that the matter comprising a protonis a superfluid.... The scaling property... follows from the fact that because a superfluid has zero viscosity there is no length scale." Chapline was not aware of our work,[66] where we arrived at the same conclusion (the absence of a scale is reflected, in our approach, in the phonon spectrum, which we had found in Ref. 48), starting from completely different experimental data. Further possible manifestations of superfluidity in deep inelastic e-p reactions were discussed nine years later by Chela-Flores (cf. below).

Also in 1975 Goble[67] starting from our results on superfluidity of hadronic matter proposed a model of pion superfluidity of *nuclear* matter. Among other things he compared in great detail the characteristics of pions in nuclear matter, which follow form his model, with those of pions in the cloud of a nucleon obtained in Refs. 48, 49 and 52.

Shortly after publishing with Ady Mann our papers on superfluidity of hadronic matter I visited Nambu in Chicago. Nambu was interested in the subject, among other things, because in

1961 he had published with Jona-Lasinio the classical papers on "Dynamical Model of Elementary Particles Based on an Analogy with Superconductivity"[68] in which they suggested that the nucleon mass arises as a self-energy of a primary fermion field, in analogy to the energy gap in superconductors, and that the pion can be regarded as a bound state of a nucleon-antinucleon pair. As far as I remember, at that time (1972?) neither Nambu nor I realized the true link between our approaches. This link became clearer (at least to Shalom and me) around 1976 when we associated in our paper[59] the superfluid phase of hadronic matter with the spontaneously broken (chiral) symmetry phase (this symmetry was present, also, in the original Nambu-Lasinio model), and where the "primary" fermions of the Nambu-Lasinio model were identified with quarks.

Nambu introduced me to Sudarshan, who happened to be at that time in Chicago, and we had an interesting chat in three, on this subject, in Sudarshan's hotel room. I remember this meeting because Nambu drove me to the hotel, where Sudarshan was staying, although the hotel was only a few minutes walking distance from the physics department of the University of Chicago. The reason was that we had to pass through an area, which was, as Nambu explained to me, one of the unsafest neighborhoods of Chicago. I had not met Sudarshan before, although we had corresponded in the sixties about the controversial issue of the pion decay asymmetry. Sudarshan was also interested in superfluidity of hadronic matter and told us that he had thought about it years before.

As we have seen the conjecture of superfluidity of hadronic matter was derived by using several analogies with He-4: both are systems of bosons with a typical Bose–Einstein distribution, both systems undergo a phase transition accompanied by a change of symmetry and in both systems we have a critical temperature and a critical velocity. But there also exists an interesting analogy between the phenomenon of superfluidity of He-4 itself and another phenomenon, that of superconductivity. In a certain sense the conduction electrons in a superconductor constitute a superfluid. To see this it is enough to realize the analogy between the hydrodynamical pressure and the electrical potential difference. According to Ohm's Law, the last quantity is proportional to the electrical resistance. In a superconductor, by definition, there is no resistance and hence no potential difference. Using the analogy with hydrodynamics it follows that in a superconductor there is no pressure "head," which means that the fluid made of electrons doesn't meet friction and is thus superfluid.

Other aspects of the analogy between superfluidity and superconductivity are: a) the energy gap, which in the case of superfluidity separates the phonon and the roton spectrum, b) in both phenomena we have the ground state made of condensates: quark and gluon condensates in superfluid hadronic matter and Cooper electron pair condensates in superconductivity.

We mentioned the link between superfluidity and confinement. There exists a link between superconductivity and confinement as well. Electron superconductors have the remarkable property that they repel external magnetic fields (the Meissner effect). In analogy, the QCD vacuum has the property that it repels color fields. Since quarks and gluons carry color, they cannot exist as free objects. The Nambu-Goldstone theorem, which states that field theories, in which symmetries are spontaneously broken, contain massless bosons ("Goldstone bosons"), is also based on the analogy with the Meissner effect in superconductivity, where the photons inside the superconductor become massive.

A further application of the analogy between superfluidity and superconductivity in strong interactions refers to quark matter. Indeed, hadronic matter is not the only particle physics system, which has superfluid properties. In the late seventies the existence of superfluidity also in quark matter was conjectured. This type of superfluidity is associated with superconductivity in quark matter at high density and low temperature. In analogy to electrical superconductivity, where electrons form Cooper pairs, quarks can form pairs, which can carry charge or color. This leads to electrical superconductivity and also to a new type of superconductivity, namely color superconductivity. In hadronic matter we have chiral superfluidity in which the condensate consists of quark-antiquark pairs, which are colorless, but break the chiral symmetry. In color superfluidity the condensate is made of pairs of quarks, which carry color and break the color symmetry. Furthermore the breaking of chiral or color symmetry in chiral or color superfluids, respectively, is analogous to the breaking of electromagnetic gauge invariance in electron superconductivity.

Related Developments

In two consecutive papers Chela-Flores[69] proposes the concept of *anisotropic* superfluidity of hadronic matter, in analogy with superfluidity of He-3. He argues that the model of Mann and Weiner is rather an analogy with He-4, because it uses pions, i.e. bosons as constituents of the nucleon. In the model of Chela-Flores the nucleon is made of elementary fermions, which form (Cooper) pairs. Such a system is analogous to the (anisotropic) superfluid A phase of He-3. A possible phenomenological feature, through which one could differentiate the two approaches, is, according to Chela-Flores, the energy gap, which is characteristic for Cooper pairing. As a matter of fact, the model of Eliezer and Weiner,[59] which supersedes that of Mann and Weiner[48,49,52] contains already fermions (more precisely quarks) as constituents. However the two approaches differ through the fact that the model of Eliezer and Weiner embodies explicitly the constituents (quarks) and the symmetries of the standard model, while that of Chela-Flores leaves the question of the nature of fermions and of symmetries open. In a subsequent paper[70] this author identifies the fermions with constituents of the nucleon (partons) and relates the disappearance of the gap above a critical temperature with the confinement phenomenon. In Ref. 71 Chela-Flores discusses possible consequences of the energy gap for the scaling behavior of the structure functions in deep inelastic lepton-nucleon scattering. The author conjectures an abrupt onset of scaling violation at very small values of the scaling variable, due to the strong correlations of the sea quarks, which form Cooper pairs.

Also, in a paper devoted to strong gravity,[72] Chela-Flores and Varela point out the correspondence between a massive charged scalar-field wave equation with a generalized Gross-Pitaevski equation for superfluidity. Based on this correspondence the authors relate the parton-parton (gluon-like) average potential strength with the overall (confining) strong-interaction strength. Hadron structure is then the result of a hadronic source in a state of superfluidity (identified with the sea quarks). This is the source of strong gravity, which curves space-time according to the strong-gravity field equations. That in turn leads to confinement of valence quarks, which carry the quantum numbers of hadrons, like electric charge, baryon number, strangeness, etc.

The concept of an outer bosonic cloud of the nucleon, which we used in 1971 to arrive at the idea of superfluidity of hadronic matter, has continued to be used also after the enthronement of QCD as the theory of strong interactions, because it reflects a non-perturbative aspect of QCD. In some recent papers Islam and collaborators[73] apply it to elastic proton-proton scattering. These authors consider the nucleon as made of an outer quark-antiquark cloud, and a core, which carries the baryonic charge. The outer cloud is responsible for the non-perturbative glancing elastic scattering at small momentum transfer, while the core, which carries the baryon quantum numbers, is a soliton-like object and determines elastic scattering at large momentum transfer (hard scattering). At some large value of momentum transfer a chiral phase transition from the nonperturbative region to the perturbative region occurs. While this approach does not explicitly use the concepts of temperature and statistical equilibrium, the similarity with the ideas outlined in Ref. 59 is striking.

Superfluidity of Hadronic Matter in Retrospective

At present most of the above-mentioned considerations are embodied in an effective field theory, which contains the symmetries of QCD. The chiral symmetry is broken in the superfluid phase by the quark condensate. The relative amount of condensate is the order parameter of the phase transition. Above the critical temperature the condensate "melts" and we are left with "normal" quark matter. The pion is a pseudo Nambu-Goldstone boson. As to Hagedorn's theory, its main message is probably the Hagedorn temperature, which is identified, today, with the critical temperature characterizing the transition from the superfluid, confinement phase to the quark-gluon deconfined phase. The exponential mass spectrum, which follows both from Hagedorn's statistical bootstrap as well as from the dual resonance model, is a very interesting and beautiful theoretical concept, but as shown in Ref. 74 it is incompatible with the existence of a quark phase. From this point of view it shares the fate of many important developments in the history of physics: it constitutes an intermediate step in the progress of science and exemplifies the Münchhausen effect.

Statistical Concepts Applied to Weak Interactions

In the academic year 1972–1973, which I spent at Imperial College, particle physics was marked among other things by the important paper due to Kirzhnits and Linde,[46] "Macroscopic Consequences of the Weinberg Model" that impressed many people in London, Shalom Eliezer and myself included. Actually, soon we found out that this happened also in the USA; S. Weinberg and L. Dolan and R. Jackiw, published in 1974 important papers inspired in part by that 1972 paper. Kirzhnits and Linde proved that for a system described by a unified theory of weak and electromagnetic interactions, a symmetry broken at low temperatures could be restored either at high temperatures or in the presence of strong, external electromagnetic fields. As a consequence of this restoration the masses of the intermediate bosons vanish and the weak interactions become long range. The authors applied their theory to large systems in thermodynamical equilibrium, in particular to the early universe. Their work implied that during the cooling process of the early universe the symmetry of the electroweak interaction was broken and took the form we see at present in the lab.

Subsequently Salam and Strathdee[75] considered in some detail the effects of external magnetic fields on the symmetry restoration.

Their calculations showed that magnetic fields of the order of 10^{16} Gauss were necessary to achieve this, while in the laboratory only fields of the order of 10^6 were available. Despite of that, Salam encouraged Butterworth and his collaborators at Imperial College to do the experiment. When I asked him why he did that, he said: "May be our calculations were wrong; let them try." To be wrong by a factor of 10^{10} is something only geniuses like Salam could afford. Of course, the results of the experiment were negative, but in hindsight I tend to agree with Salam: Physics is an experimental science and as such has always been good for surprises.

Around that time I landed in Marburg and Shalom moved to the University of Massachusetts in Amherst where I visited him. After that we met in Los Alamos where I spent a sabbatical semester and later on Shalom came for a few months to Marburg with a Humboldt fellowship. We used our overlap to study changes of symmetries in high-energy particle reactions at high temperatures. Our work differed from that mentioned above in two respects: we were, in the beginning, concerned with strong interactions and for us the microscopic systems met in particle reactions were (almost) as good thermodynamical systems, as the universe. In other words, the usual *heat bath* was provided by part of the very system under consideration.[b] To emphasize this we introduced in our 1976 paper on phase transitions in strong interactions[59] the concept of *internal temperature*, to be compared with that of external temperature, used in papers dealing with field theories at finite temperature. However, while the use of an internal temperature, measured by the average transverse momentum of produced secondaries, had been limited until then to strong interactions, in the mid 1970s Shalom and I started to ask us why this approach could not be extended to weak interactions. Since at high temperatures weak interactions become strong, the justification of a statistical approach, used for strong interactions, would apply also for weak interactions.

---

[b]Actually, this is also true for the early universe, although it is usually not stated explicitly. Indeed, in the early universe the "heat bath" was also part of the system, the universe comprising, by definition, everything.

The difficulty this extension met seemed at a first look rather of experimental nature: The temperatures involved in weak interaction (10^3 GeV) were by four orders of magnitude higher than those in strong interactions, which implied a similar increase in the transverse momenta of produced particles. Still, the *energy densities* produced in weak interactions and hence the associated temperatures were also correspondingly higher, because the characteristic distances (and volumes) were accordingly smaller, the characteristic distance being determined by the inverse of the mass of the exchanged particles. While for strong interactions the characteristic mass is that of the pion, for weak interactions it is that of the intermediate boson. In 1978 when we decided to write our paper "Changes of symmetry in weak interactions at large transverse momenta"[76] the intermediate bosons had not yet been discovered, but the order of magnitude of their masses could be guessed, among other things from the magnitude of the weak coupling constant.[77] In Ref. 76 we considered the possibility that the violations of symmetries of weak interactions that appear in the standard model, in particular parity and charge conjugation, are due to spontaneous breakdown. We pointed out that in this case the symmetries would be restored at high temperatures and this could be seen in the laboratory at high energies and large transverse momenta. In particular, since parity violation implies that muons produced in weak interactions are polarized, one should observe a depolarization of these particles with increasing transverse momentum. A very simple formula for this striking effect in terms of the mass of the Higgs and the scalar coupling constant was derived. Although these two parameters were not known in 1978 and, as a matter of fact, are not known even today, an order of magnitude estimate showed that transverse momenta higher than 300 GeV would be necessary to see experimentally this effect. These large momenta are not available even at present, but the LHC accelerator under construction at CERN might get nearer to this goal.

Photographs

The story of the photographs referring to the years before 1969 and that of my ghetto ID card is in itself perhaps worth mentioning. During the ghetto years, we lost most of our belongings including family photos. When my father and I moved in 1945 from Czernowitz to Romania, we had with us only the ghetto ID cards and my father's university diplomas. These I left in Bucharest in my "Institute" at Labirint 28 together with my diplomas and they were recovered by the mother of my fiancée Uli after Uli and I had managed to cross the Iron Curtain. As for the family photos, two of these were recovered by my cousin Rosi Shteyn. Before being deported in 1941 to Siberia, she had left them with an Ukrainian friend in Storojinets whom she met again in the 1950's when she returned from deportation. My parents' wedding photograph I found in the 1970's at the French branch of the Weiner family.

Fig. 1. My parents' wedding, around 1928.

Fig. 2. The author at the age of two.

Nr. 1597/W.227/1942 OFICIUL JUDEȚEAN AL EVREILOR CERNĂUȚI

Carte de identitate

Evreul capul de familie / membru de familie Weiner Richard

strada Petre Liciu Nr. 8

1) Posedă autorizația și declar. de ședere în Cernăuți Nr. 1597/1941

2) Posedă adeverință dela recensământul evreilor Nr 24185/1942

3) Posedă buletinul de înscriere la biroul de populație Nr. 17802/194

Semnătura posesorului Weiner Richard

Pierderea carnetului de față atrage anularea autorizației de ședere în Cernăuți.

Delegatul Poliției, Delegatul C. E. R., Delegatul Guvern. Bucovina,

Fig. 3. The author’s ghetto ID card.

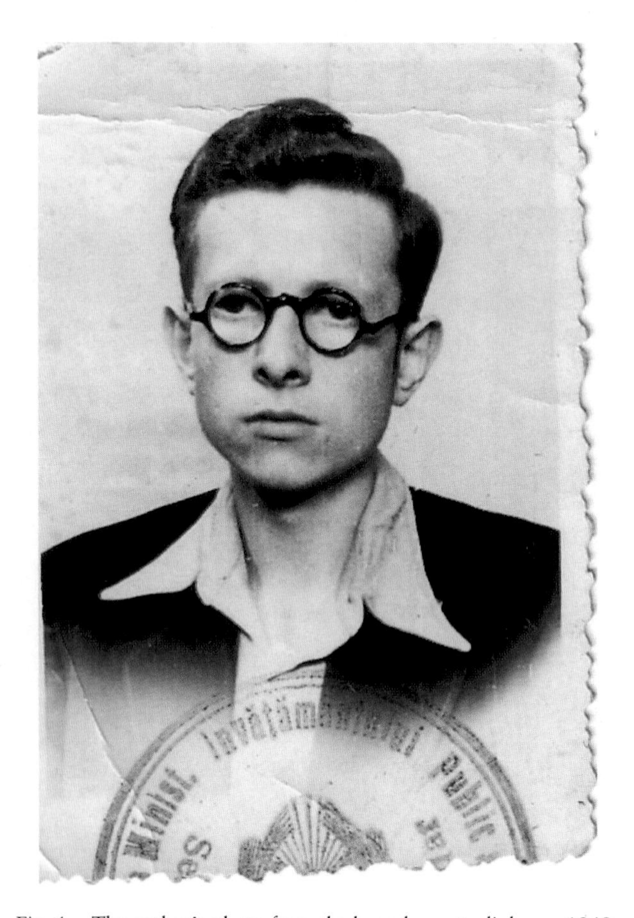

Fig. 4. The author's photo from the baccalaureate diploma, 1948.

Fig. 5. Uli, the author's future wife, 1968.

Fig. 6. The author with his uncle Muniu and Uli at the Bratislava railway station on September 24, 1969.

CERN

ORGANISATION EUROPÉENNE
POUR LA RECHERCHE NUCLÉAIRE
GENÈVE

ATTESTATION

Fig. 7. The author's CERN ID, 1969.

Fig. 8. From right to left: Peter Carruthers, Uli, Ed Shuryak, Evgenii Feinberg and the author, in the author's house in Marburg, during the LESIP I meeting, 1984.

Fig. 9. From left to right: Gerald Fowler, an unknown lady and Gerald's wife Margaret, during the LESIP I meeting, 1984.

Fig. 10. With Minoru Biyajima and an unknown person, in China, during the Shandong meeting, 1987.

Fig. 11. Some members of the Marburg group, from left to right: (first row) Herbert Neuberger, Sibaji Raha, the author, Grzegorc Wilk; (second row) Fred Pottag, Udo Ornik, Fernando Navarra, Michael Plümer, 1988.

Fig. 12. In Japan, with Naohiko Masuda, 1991.

Fig. 13. Erwin Friedlander (second from right) and the author in 1992, in front of the author's house in Marburg. The other two people are Alec Leicand, a common friend from Sao Paulo, and Daniel Leicand, his son.

Fig. 14. From left to right: (seated) E. Papp, Igor Andreev, Leonid Razumov; (standing) the author, in the author's house in Marburg, 1995.

Fig. 15. Paris, 2000, in front of Théatre de Chatelet: an unknown lady, the author's daughter Diana, Uli's mother Nana, the author and Uli.

Fig. 16. From right to left: Shalom Eliezer, Shalom's wife Yaffa and the author, at Versaille, around 2000.

PART II

Settling Years (1974–present)

Section V

Marburg; Hot Spots

14

Professor at the Philipps University of Marburg

The position of a professor in Theoretical Physics at the Philipps University of Marburg had been advertized in 1973 in a German newspaper. It was mentioned that it implied the status of a "civil servant for life" (*Beamte auf Lebenszeit*), which for somebody who had changed only in the last four years four countries and as many visiting positions, not to mention the previous four changes of allegiance during the period 1940–1945, was a very attractive option. The fact that I was already 43 and had a family played an equally important role in my considerations. That is why I applied for this position. Shortly after that, I was invited to Marburg to give a talk and was offered immediately the position. Before my visit to Marburg I had not known anybody there even by name, except Ludwig, whose work on foundations of quantum mechanics was somewhat familiar to me, and Wilhelm Walcher, an experimentalist, one of the 18 German atomic physicists who had signed in 1957 an important declaration stating that Germany would not build an atomic bomb. Paul with whom I discussed this offer and who was somehow familiar with the situation in Marburg told me that although the position was much below my qualifications (it was a H2 professorship in a scheme which contained two higher degrees) I would be able to continue my research without too heavy teaching duties, since the department had about 40 professors, quite a large number for a medium-sized university like that of Marburg, and many of these professors preferred to teach rather than to do research. Besides that, it would be just a starting position, wherefrom I could look for something better. That is why I accepted this position. The chairman of the Physics Department in Marburg had even offered to help me bridge the time until my official appointment by a visiting position, but that was not necessary, since I had already a visiting appointment at DESY. Only a few months after joining the University of Marburg I was offered another higher rank professorship at the University of Duisburg, which was a new university and where I was supposed to contribute to the development of the physics department. I did not accept this new offer among other things just because of that last circumstance: I was afraid that I would have to spend most of my time with organizational matters, which horrified me. Actually, I got a foretaste of what that might mean after I had been nominated into the council of the Physics Department at Marburg University. The council met almost weekly and a meeting took several hours, quite often without leading to anything concrete. This was in part due to the over-democratization of the organization of the university, a consequence of the 1968 student movements. While before 1968 the chairman of the department and the "ordinary" professors decided everything, without consulting anybody else, not even their colleagues — the lower rank "extraordinary" professors — after that date even students and administrative staff had a heavy say, and that not only in matters that directly concerned them, but also in pure scientific and research problems. Thus soon after my arrival the department had to decide whether to fill in a professorship vacancy with a nuclear or a particle physics experimentalist. This decision, like any other, was taken by people many of whom had no idea what the difference between these two fields was. On the other hand the reform of the university organization maintained in most cases some privileges of ordinary professors, like research associates and assistants paid by the university. These important details were unknown to me at the moment when I accepted the position: I was accustomed with the department structure of American universities, where research associates were free to decide what and with whom they would like to work and therefore I felt quite lost. Eventually I managed to find a collaborator among these assistants, but that only because the professor he was associated with was not interested in research. Actually, according to an old German proverb the difference between ordinary and extraordinary professors was that the former never did something extraordinary. Another peculiarity of the German university system, which distinguishes it from those of other countries and which I had not known before my appointment was that this system did not allow promotion at the same university. This rule, albeit unwritten, had been introduced to prevent favoritism. A further unpleasant surprise was the way the yearly budget of the department was handled. I wouldn't have dreamed of, that even the expenses for paper, Xerox and post stamps, not to mention telephone conversations, were strictly controlled according to a key, which had nothing in common with the real needs of research. Moreover, unlike travel expenses, these could not be supported from research projects funds. The post stamp problem concerned me in particular, because I was the only one who had international collaborations and the only one who was sending out preprints. To discourage possible followers of my example, the expenses for post stamps for each group were listed yearly in the budget report of the department and this procedure continued to be applied even twenty years after my arrival in Marburg, although I used to pay these expenses from my own pocket.

Fortunately this was not the whole story of my Marburg experience. Actually, after a while I started to appreciate the atmosphere and cosiness of Marburg, a typical German university town, where there was almost nothing except the university (there is a saying: there are cities which have universities; Marburg *is* a university). Duisburg, on the other hand, was a highly industrialized big city in the Ruhr region, which was more famous for its pollution. Furthermore, the University of Marburg, as one of the oldest European universities, actually the oldest protestant university, had quite a fame: names like Bunsen, Otto Hahn, Lomonosov, Pasternak, Cohen, Heidegger, Hannah Ahrendt, who taught or studied there, were associated with it.

The circumstances leading to its establishment are quite amusing: it was founded in 1527 by the 23-year-old landgrave Philipp the Magnanimous, the Prince of the land of Hessen, to educate "learned, able, and God-fearing persons, preachers, and officials for Christian benefit and the good of the common land" and two years after its foundation, in 1529, Philipp invited to Marburg Luther and Zwingli, the two most important protestant theologians of that time, to settle in his presence their famous dispute: whether the red wine Christians drink on Easter is the symbol of Christ's blood, as Zwingli sustained, or really his blood, as Luther thought. Actually Philipp was not such a pious Christian, as this act of university foundation would suggest: he used his relations to obtain from Luther (and Melanchton, another leading theologian) the permission to marry a second wife and to live in bigamy. Besides the leading theological faculty, the university had from the very beginning also faculties for jurisprudence, medicine, and philosophy, but because of the particular circumstances of its foundation, the theology department has been for a long time one of the most famous theology departments in the world.

Last but not least another reason why I remained in Marburg until my official retirement in 1995 was that almost immediately Uli also found at the university a permanent position as a chemist, albeit in a different field than her initial specialization: she became after getting her PhD quite a successful and appreciated neurochemist.

Citizenship

One of the things, which preoccupied me during my stay in Bonn was the issue of citizenship, or rather that of a Western passport, which would allow me to travel outside Germany, without having to apply for each country for an entry visa, a rather lengthy and complicated procedure. As mentioned before, this was one of the reasons why I decided not to remain in Switzerland, where it would have taken a very long time to get citizenship. To my surprise this turned up not to be as simple as I thought, in Germany as well. It was another piece of evidence for the discrepancy between the image of Germany and what was behind this image. Mislead by that image I thought it unnecessary to accept the help offered by Rollnik in handling this issue. That turned out to be a mistake and in hindsight I believe that he was probably aware of the obstacles I would meet, although the law established soon after the founding of the Federal Republic of Germany was quite clear in this respect: people of German background from Central and Eastern Europe, in particular those from Romania and the Soviet Union, had the right to get German citizenship immediately.

Anyway, my six-month stay in Bonn was not sufficient to solve this problem. Among other things, a letter addressed to me by the German authorities on this issue after I left for Bloomington, did not reach me in *legal time* so that after my return to Germany the authorities declined to reopen my file and I would have to wait the usual 5–8 years. (A similar delay in forwarding the mail — the secretary had probably overlooked the need to specify AIR MAIL on the envelope — made me miss the invitation to the University of Zürich mentioned before).

Of course, once appointed "Beamte auf Lebenszeit" of the Land of Hessen I was entitled to get citizenship within a shorter time than other immigrants, but this would still be a matter of years. However, I was counting on the law, which *entitled* people of German background to German citizenship. What I did not know at that moment was that the proof of German background in the particular case of Jews from Romania was not straightforward: The Romanian passports or birth certificates did not mention the ethnicity or nationality. Therefore the proof consisted of various elements like knowledge of German language,

German name and certificates proving membership in German cultural or sport organizations. For Jews who after 1933 were not accepted anymore in these organizations the situation was more complicated and raised among other things the question of principle: were these people Germans by nationality and Jews by religious faith or were they Jews by nationality *and* faith. And what about those who, like my father, were freethinkers and did not consider themselves belonging to any religious faith? In a truly secular state like France, faith is a matter of private opinion and has no bearing on the issue of nationality or citizenship. Not so in Germany. The complexity of the German approach to the problem of citizenship for Jews from Romania was enhanced by the fact that the decision about who is, or who was what, had to be taken by immigrant organizations with semi-official character, which were renowned for their revanchist, if not straightforward anti-Semitic character, being the successors of the German ethnical organizations in Central Europe. Indeed, it is a historical fact that almost all ethnical German organizations of Czechoslovakia, Hungary, Poland and Romania were consequent executors of the policy of Nazi-Germany and many, if not the majority, of their members, perhaps even more than their countrymen in mainland Germany, were ardent admirers of Hitler. In Czechoslovakia, for instance, this manifested itself, among other things, through the enthusiasm with which the Sudeten-Germans welcomed the annexation of their province, Sudetenland, by Nazi-Germany. Another symptom of this allegiance to Nazi ideals was a ferocious anti-Semitism. This explains why in the Bukowina, even before World War II, when still a Romanian province, Jews were ousted from German cultural or sport organizations, although they had founded these organizations and constituted for decades their backbone. The significance and bitter irony of this act of segregation can be better judged by reminding that in the Bukowina in particular, German culture had been represented for many decades by people of Jewish origin, who considered themselves as belonging to the Austrian-German culture and who not only spoke, almost exclusively, German, but who had created important works of German art and literature. It is also a sad irony of history that this cultural activity continued in the Bukowina outside the ethnic German organizations, even at a later time, when in the Third Reich Jews had been eliminated from the German cultural life and their physical elimination was in preparation, not to mention that the very notion of culture was considered by some Nazi leaders as despicable.

Anyway, in the concrete citizenship case of Jews from the Bukowina, being of German culture was not considered sufficient and had to be supplemented by a proof that the person in cause had declared to be German during the national census, which took place in Romania in 1930. Since I was born that year, the proof applied to my parents. Moreover, since these census declarations were anonymous, I was asked to find witnesses who would confirm that my parents declared themselves of German nationality forty years ago, and that in a country which no longer existed and where most people had been deported and exterminated. The absurdity of this requirement, which, incidentally, I eventually managed to fulfil, makes it difficult to avoid the impression that the ethnicity clause based on membership in German organizations or on census statements had been specially invented to prevent Jews of getting German citizenship. When I told this story to Kurt Schöpflin he was not only amazed but furious. In some sense he felt responsible for "his" new and democratic Germany, and, like many others he saw in that more than a manifestation of bureaucracy. Indeed, a short note addressed by him to the Prime Minister of Nordrhein-Westfalen, Heinz Kühn who had been himself an emigrant during the war and who, like Kurt, had been a member of the social-democrat party for more than fifty years, made wonders: within a few weeks my citizenship problem was solved.

What I resented most during my activity at Marburg was the relative scientific isolation. Except Mathias Weström, a research associate of Grawert, I did not find among the physicists of the department any collaborator, although immediately after arriving there I tried to establish with the professors in the Physics Department a closer scientific contact. One of the reasons for this communication difficulty was the fact that I was phenomenologically oriented while my colleagues in theoretical physics, almost all former students of Ludwig, were mostly interested in problems of foundations of quantum mechanics. When I met for the first time Ludwig he made a point in emphasising that for him relativistic quantum mechanics "did not exist." Actually, that was in line with the fact that von Neumann had limited his formulation of quantum mechanics to the non-relativistic case and that a relativistic extension of it could not be achieved but in field theory. Field theory, however, was plagued by the renormalization problems and was therefore unreliable. Von-Neumann's and Ludwig's views were in a certain sense contradicted by the less axiomatic, but much more productive point of view according to which Dirac's theory was the relativistic extension of non-relativistic quantum mechanics. However this discussion reminded me of my own "experience" with these fundamental problems and in particular the problem of interpretation of time in quantum physics (cf. chapter 5). Anyway, at the moment I joined Marburg particle physics, not to mention high-energy phenomenology did not even exist there. In a certain sense this situation reminds of that in post-war France and which is described in the memoirs of some French physicists like Abragam and Friedel. In French universities until 1968 quantum mechanics did not make part of the obligatory curriculum of physics. (From this point of view at least the situation was better even in Romania). As far as I can judge, the reasons for this underdevelopment in Germany and France were however different. In France it was the strong influence, some would say the dictatorship, of Louis de Broglie who had prevented any contact with the new developments in field theory. In Germany it was probably in part due to the haemorrhage produced by the loss of Jewish or anti-Nazi physicists. Common to both these countries was however the loss of contact with the USA and Britain, where the main new developments of quantum field theory took place. To improve this situation gifted students from France and (to a lesser extent) from Western Germany could and did obtain fellowships that permitted them to go abroad, in particular to Britain and the USA, where they could actively follow and participate in the new developments.

Stranded in Marburg I attempted to fight my initial isolation there by frequent travels, mostly to CERN, the United States and Britain, and by inviting my former collaborators like Shalom Eliezer and Ady Mann to Marburg. A few years later I started an intensive and long lasting collaboration with Gerald Fowler, whom I had met during my London stay, and Erwin Friedlander, who came in 1976 out of the cold and for whom I managed to find in Marburg a first harbor, before he settled at Lawrence Berkeley Lab, and with Peter Carruthers. Both Gerald and Erwin spent sabbatical years in Marburg and I spent my sabbaticals in Berkeley and Los Alamos. The collaboration with Peter Carruthers also amounted among other things in the organization of the LESIP meetings, which I had initiated in 1984 in Germany. Later on Igor Andreev from the Moscow Lebedev Institute also spent a sabbatical in Marburg.

When I arrived in Marburg at the beginning of 1974 I was "carrying" with me an idea I had gotten during a trip from London to Israel and which I had started to pursue at Imperial College and then at DESY. As a matter of fact it was not the first or the last time that I got inspiration on a trip and I remember that quite a few times I worked in the plane or at airports with Erwin Friedlander or Peter Carruthers; I am aware that this is a common phenomenon among physicists at least. The idea I was "pregnant" with had to do of course with what I had been working on in the previous years.

15

Hot Spots in "Elementary" Particles and in Nuclei

Propagation of Heat in Hadronic Matter

The isomeric shift and superfluidity of hadronic matter are effects in which the finite size of the system play a major role. Another effect of this kind and which is also based on an analogy is the hot spot effect. This effect refers, as its name already suggests, to a space region, which is locally heated and it is a common phenomenon in many chapters of macroscopic physics, among others condensed matter and astrophysics. Thus e.g. most recently hot spots have been reported in accreting pulsars.[78] Being interested in finite size effects in microscopic physics I started to wonder whether, by analogy, one wouldn't see this effect in this field, too. While in the isomeric shift it was the finite size of the nucleus, with superfluidity of hadronic matter and the mesonic cloud analogy I started to be preoccupied with the finite size of what was considered, at that moment, an "elementary" particle, the nucleon. However, although the hot spot analogy I will discuss here referred initially to the nucleon, it was later extended to nuclei as well.

The hot spot effect and superfluidity of hadronic matter have also another common feature: both are based on statistical concepts.

Indeed, superfluidity is a state of matter and this presupposes global statistical equilibrium. The hot spot is also a statistical concept, albeit less "demanding," and therefore more general: it assumes local equilibrium only. Actually, local equilibrium is the precursor of global equilibrium and the hot spot effect can be used to determine how fast, if at all, the transition from local to global equilibrium takes place. That this transition does not always happen follows from the simple fact that the duration of a strong interaction reaction is quite short (of the order of 10^{-22}–10^{-23}seconds) and the propagation of "heat" i.e. of the excitation through the body of the system takes a finite time, which is determined by the heat conductivity of the matter the system is made of.

Indications of the transition between local and global equilibrium in particle physics started to emerge in the 1960s and early 1970s. In high-energy strong interactions equilibrium is not complete. In particle reactions with the increase of laboratory energy one observes that the transverse momenta of produced particles have a tail, which deviates from the single exponential Boltzmann spectrum, characteristic for global equilibrium. The slope or the effective temperature of this transverse momentum tail increases with increasing energy. Feinberg and others interpreted these large transverse momenta as being due to particles, which "leak" out before equilibrium is reached. Similar observations had been made in nuclear reactions and were also attributed to pre-equilibrium effects. This interpretation suggested that the equilibrium is neither instantaneous, nor global, but rather local in space and time. However, in the absence of a theory of strong interactions applicable to non-perturbative processes this conclusion could hardly be proven theoretically. That is why in the early 1970s I was thinking from time to time about a possible experimental test of this hypothesis. And one day, out of a sudden, I realized that the finite size of the nucleon, itself, could serve as a tool in such a test. Strangely enough, I remember even today, after more than 30 years, how I got this idea. It was sometime in spring 1973 and I was flying from London to Israel. The plane was approaching the Ben-Gurion airport and I was looking out of the window trying to get a glimpse of the "promised land" (it was my first trip to Israel). I couldn't see much because it was a rather foggy day and also because my seat was on a wing. Instead of land I saw bits of clouds surrounding and rolling over the wings. This slide of the plane through the mist made me think about the friction between the wing and the air and how the resistance the clouds opposed to the plane influenced its flight. Despite the low density of the air at high velocities this effect is important and explains why the shapes of aeroplanes and wings are designed "aerodynamically," i.e. to minimize the friction. It also explains the heating of a rocket or satellite when it returns into the earth's atmosphere. The friction between the rapidly moving plane or rocket and the atmosphere is essentially a surface effect and this reminded me of the mesonic cloud concept, which emerged from the study of peripheral reactions. It is conceivable that the word cloud, which appears in both phenomena, triggered this analogy in my mind. And at this moment, what had been buried in my subconscious for several years, started to come out into the open. I had been asking myself for a long time what is the physical difference between peripheral collisions, which contribute to the bulk of the cross section, and central collisions, which are very rare. The word *peripheral* suggests that the reaction occurs on the surface of the particle (which obviously makes sense only if the particle has a finite size). Often instead of peripheral the term grazing is used and this brought me back to what I was seeing outside. It came to my mind that in analogy to the friction between the wing and the air, a grazing collision between two particles could also be viewed as a surface effect. This lead me to the idea of the hot spot effect in particle physics: I conjectured that in certain peripheral collisions of hadrons — I had in mind strong interactions — the interaction between the two particles is localized on a small region of the surface of the particles, which is consistent with the fact that strong interactions are short ranged. In analogy to the aerodynamical effect, this spot gets "heated" and here from the name of the effect.

To see whether this picture had something to do with reality I looked for observable consequences. The consequence I came up with was the prediction of an asymmetry in certain reactions, in particular in pion production reactions. Indeed, the hot spot is an excitation of the particle, which has to be followed by a corresponding de-excitation. Such a de-excitation can occur through particle emission. If one defines an "above" and a "below" in terms of the projectile,[79] then above the hot spot there is vacuum and the emission can take place without impediment. A detector facing the hot spot from "above" will register this emission. "Below" the hot spot, however, there is the body of the particle, its "interior." Here at least a fraction of the emitted particles will be absorbed, so that a detector facing the hot spot from below will register less emission or none. This leads to an "up-down" asymmetry (see Figure 1).

The observability of the effect depended crucially on whether there were not other competing mechanisms that contributed to the de-excitation of the target (the same considerations hold, of course, also for the projectile, which can also develop a hot spot). Such a competing mechanism was, obviously, the cooling of the hot spot through heat propagation within the body of the target. This propagation depended on the (unknown) heat conductivity of hadronic matter. If the heat conductivity were large, thermal equilibration would take place rapidly and there would be no time for preferential radiation from the hot spot into the upper hemisphere and the asymmetry effect would not be observable. I had, however, reasons to believe that the effective heat conductivity of hadronic matter was small. This happens for nuclear matter (cf. below) and also for superfluids below the critical temperature and my pending work with Ady Mann on superfluidity of hadronic matter was probably in my subconscious. But even if one took the stand that heat conductivity of hadronic matter was unknown, the search for the asymmetry effect would become equivalent to the measurement, for the first time, of that property of hadronic matter. This fact convinced me (and others) that the investigation of the asymmetry was worthwhile both form the theoretical and the experimental point of view.

Arrived in Israel I discussed this issue, among others, with Jacques Goldberg, the leader of the high-energy experimental group

The 'hot spot' effect, in a particle or nucleus, where a glancing collision creates a local region of excitation. Results from three recent experiments indicate that such effects have been seen and could provide an additional way of learning some particle and nuclear properties

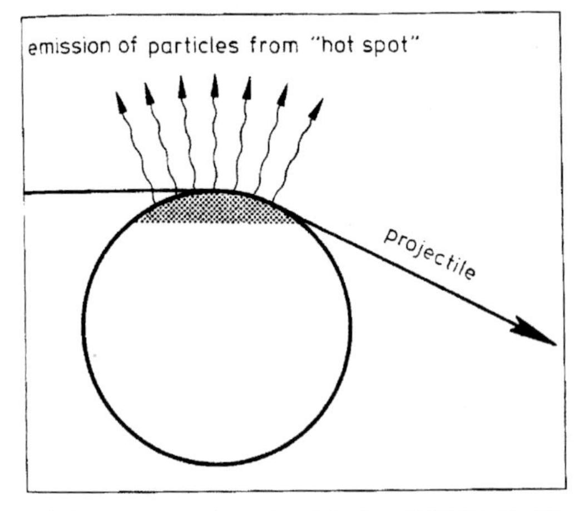

Hot spots discussed at Bonn

Is it reasonable to talk about a fraction of a particle or nucleus? Is there an 'up' and 'down' in such microscopic systems? A short time ago such questions appeared meaningless or, at least, highly speculative but this is changing.

By reformulating the questions — Can we localize excitations in a parti- or nucleus? Is such a localization observable? — we begin to see positive answers from three independent experiments performed recently in Europe, Israel and Japan. The importance of this development for nuclear and particle physics is that it becomes feasible not only to follow the space-time evolution of a reaction but also to gain new information about nuclear matter (especially heat conductivity) which was not previously accessible to experiment.

The starting point of these developments were theoretical studies (particularly by R.M. Weiner) at the University of Marburg from 1974. New effects were predicted as a consequence of local excitations called 'hot spots'. These studies were supported in part by DESY, Los Alamos and Deutsche Forschungsgemeinschaft.

When two particles or nuclei have a glancing collision, it is expected that at first only their surfaces will be excited. Theoretical considerations suggest that this excitation will propagate rather slowly into the interior of the system and any emission of particles should therefore occur before the excitation has spread over the whole system. Since the excitation is only local, this emission takes place from a limited region which leads to a space asymmetry in the emission of particles.

Such an asymmetry can be seen in a coincidence experiment which detects one of the scattered particles and an emitted particle. Observing the scattered particle gives the asymmetry direction and, from the magnitude of the asymmetry, interesting information about transport properties in nuclear matter as well as about the size of the excited region can be obtained. In this way, the concept 'fraction of a particle' gets some scientific meaning.

The recent experiments involved particles (Israel Institute of Technology) and nuclear systems (Max-Planck-Institut für Kernphysik, Heidelberg/Centre de Recherches Nucléaires Strasbourg/Institute of Physical and Chemical Research, Wako-Shi). In all three experiments the asymmetries are consistent with the hot spot mechanism.

Although a complete understanding of hot spots needs much more elaborate experimental and theoretical investigation, at present this kind of localization of excitations seems the simplest explanation of the observations. This was a conclusion from an international workshop organized by Bonn University from 15-17 November. Further experiments are being performed at various institutes including Berkeley and Darmstadt.

GSI Darmstadt has recently submitted a proposal for the construction of a high energy heavy ion accelerator and a similar project is under study at Berkeley. Such high energies are necessary to investigate hot spots with all their consequences. One of these consequences is the possibility of achieving extremely high temperatures (trillions of degrees!) in very small volumes. At present the highest temperature obtained in the laboratory is a million times smaller.

Fig. 1. (From CERN Courier 19 (1979) 24).

at Technion, who manifested interest in looking for this effect. And, what was equally important, Jacques was a member of the Rutherford-Saclay-Ecole Polytechnique collaboration and he thought that some data taken by this collaboration could be used for this purpose.

I, myself, after returning to London, continued to work on the theory of the effect. The task of the theory was first of all to calculate certain physical quantities in terms of heat conductivity, so that experimental determinations of these quantities could be used for the measurement of this transport coefficient. The obvious candidate for such a quantity was the temperature as a function of space and time. Initially the temperature is concentrated in the hot spot, the rest of the system being at zero temperature. Then from the hot spot the temperature spreads over the body of the nucleon. I assumed that in a first approximation this propagation of heat or temperature is described by the classical relativistic diffusion equation, which had to be solved with appropriate boundary and initial conditions. Luckily enough, there existed in the literature an exact analytical solution of the diffusion equation, so that I could calculate the temperature field and estimate, as a function of heat conductivity, the "up-down" asymmetry in certain physical observables, like numbers of secondary particles produced, their average transverse momentum and their masses. As was to be expected, the smaller the heat conductivity, the bigger the asymmetry was.

In the mean time I had moved from Imperial College to DESY in Hamburg. Here I wrote up a short version of my results, published it as a DESY preprint and submitted it in December 1973 to *Physical Review* Letters, where it appeared three months later[80] under the title "Asymmetry in peripheral production processes." In 1976, during a sabbatical at Los Alamos, I published a more detailed version in *Physical Review*, entitled "Propagation of heat in hadronic matter".[81]

A few years later Jacques Goldberg published his results.[82] I quote here the abstract of his paper: "Properly selected pions peripherally produced in the reaction $K^- p \to K^- p\, \pi^+ \pi^-$ at 14 GeV/c exhibit an asymmetry in a particular reference frame designed to check a conjecture by Weiner that such pions may be evaporated from an excited hadron before thermodynamical equilibrium is reached. This is the first observation of localization in hadronic interactions."

Before and after publishing my first paper on hot spots I had several discussions about this topic, in particular with people in London, at DESY and at CERN. At DESY, the home of the electron synchrotron, I was, not surprisingly, asked, (among others by Schildknecht and Walsh), why not look also for hot spots produced in deep inelastic electron scattering in the interior of the nucleon. If I remember correctly, this question was asked also by the referee of *Physical Review Letters*, who acknowledged that the issue of hot spots on the surface of the nucleon was "enticing." My answer was that in deep inelastic scattering it would be much more difficult, if not impossible, to study the processes of preequilibrium and heat propagation.[83] Elliot Leader had asked another, quite pertinent question: Isn't it conceivable that in peripheral collisions the target begins to rotate and in this way the asymmetry is washed out? I answered this question in the negative in a footnote in the *Physical Review Letter* (PRL) and I forgot it completely until a recent article in a newspaper brought it back to my mind. Since this issue is of more general interest I mention it in the following. In the PRL I invoked two reasons why rotational effects were not expected to influence the asymmetry: Even if a rotation would occur, due to the smallness of the momentum transfer it would be too slow to destroy the asymmetry. Secondly, if the nucleon is considered a composite system, then effects of rotation could lead to observable effects only if the nucleon is deformed. And then I continued: "There is no evidence (so far) for non-spherical shape of particles." This last sentence has remained valid until quite recently when I read in *Le Figaro* of April 9, 2003 that some new experimental data from the electron accelerator at Jefferson Laboratory could challenge this belief. Soon after that a paper on that topic by Miller[84] appeared. In essence Miller argues that for high momentum quarks the shape of the nucleon resembles that of a peanut or of a bagel. Although this fact, if confirmed, does not apply to the hot spot effect, because it refers only to high momentum transfer, I found it amusing.

I also corresponded with Feinberg from the Lebedev Institute in Moscow, whom I had sent the prepint and asked for his opinion, because I considered him the leading expert on the application of statistical methods to particle physics in the Soviet Union. (As acknowledged by Landau in his famous paper on the application of hydrodynamics to multiparticle production in high-energy physics, Feinberg had played a key role in the development of this approach). Feinberg suggested that in the hot spot effect, besides heat transport one should also consider shock waves. Although I did not quite agree with this point, since I thought that the momentum transfer was too small for shock waves to play a role, somehow this suggestion germinated in my subconscious, because a few years later I embarked with Masuda on a hydrodynamical project, which resulted in the solution of the equations of hydrodynamics for high-energy proton nucleus collisions. Before that, however, I started a collaboration with Mathias Weström, in which we investigated the hot spot phenomenon in nuclei.[85]

Hot Spots in Nuclei

The idea of a local excitation on the surface of the nucleon heavily relies on the assumption that this particle has finite dimensions. Moreover, the magnitude of the "up-down" asymmetry increases with the size of the system, since in a bigger system it has more time to develop. While the electron proton scattering experiments had shown beyond any doubt that the nucleon had a finite size, it was a-priori not clear whether this size was big enough to see the hot spot effect. There existed, however, another strongly interacting system of much larger size, albeit still in the microcosm, where the effect could be looked for, the atomic nucleus. Therefore in the 1974 paper where the idea of a local excitation on the surface of the nucleon was introduced, it was mentioned that "this effect should be enhanced if nuclei were used as targets, since the finite dimensions play here a more important part."

Still, from my point of view the conjecture that hot spots should be seen also in nuclear reactions was based on simple analogy with the particle physics case, although in hindsight and as will be explained below, one might reverse the argument and claim that the analogy should start with nuclei. Anyway, the analogy between hot spots in particles and in nuclei turned out to be quite fruitful, as it was followed by an avalanche of experimental and theoretical investigations, which, eventually, established the hot spot model as an accepted model of nuclear reactions. The experimental studies included, besides tests of the localization of the heated region, also rather sophisticated tests of its equilibration through polarization measurements of protons[86] and gamma rays.[87] The phenomena of hot spots, heat conduction and preequilibrium play also an important part in high-energy heavy ion reactions and in the search for the phase transition to quark matter.[88]

The success of the hot spot model and the implications it had for the assumption of local equilibrium was also one of the motives why I organized in 1984 the meeting on "Local Equilibrium in Strong Interaction Physics" (LESIP) (cf. chapter 17).

That the idea of hot spots was rather easily accepted in nuclear physics is due to the fact that statistical concepts were more established in this domain. This is due first of all to the pioneering work of Bohr, Bethe and Weisskopf in the late 1930s, which was based on the notion of statistical equilibrium, but also to more recent work in the seventies, when the concept of preequilibrium was applied to nuclear reactions. In particle physics the situation was different, not only because statistical concepts were considered by some physicists "heretical," but also because hot spots raised the issue whether one can talk about a fraction of a particle, about an "up" and "down" in such a microscopic system. This explains in part why the experimental evidence for hot spots in the nucleon made also headlines in the CERN Courier.

Meeting Bethe

Making contact again with nuclear physics after an interruption of 15 years, I had a surprise: I found out that in 1938 Bethe had already thought about hot spots in low energy nuclear reactions. In an abstract of an American Physical Society meeting published in the *Physical Review*,[89] although not mentioning the words "hot spot" he sketched the essence of the idea and estimated the probability of this effect in terms of heat conductivity of nuclear matter. Bethe's approach referred to low energy central nuclear collisions and therefore predicted a forward-backward asymmetry, while the effect considered by myself applied to high-energy peripheral particle reactions and predicted an up-down asymmetry. Still, the physics is the same and it is amazing that it took almost 40 years until this idea was resurrected.

I personally met Bethe for the first time in the seventies at a meeting of the American Physical Society (APS), which took place in Los Alamos. I told him about my "rediscovery" and apologized for not having quoted him in my first papers on this subject. Bethe's reaction was typical for a scientist of his rank: "How could you? It was published twenty years ago and it was just an abstract."

That statistical concepts and ideas about preequilibrium were simmering already in the late thirties follows also from the fact that in 1938 Tomonaga, at that time a PhD student of Heisenberg in Leipzig calculated for the first time the coefficient of heat conductivity in nuclear matter and found that it was low. That justified the expectation that hot spots in nuclei would be observable. I don't know whether Bethe, when he presented his talk at the 1938 APS meeting, was familiar with this result.

As to the APS meeting in the 1970s, I was not scheduled to talk at that meeting. However Bethe, who was chairing a session of the meeting and who was interested to learn that the analogy of hot spots did not stop at the nuclear level, but applied to particles, too, gave me a few minutes at the beginning of his session to present my theoretical calculations as well as some preliminary experimental results of Jacques Goldberg, which seemed to confirm the existence of the effect for nucleons.

Although this was my first personal meeting with Bethe, I, like most physicists, was familiar with many of his classical works, in particular with the elucidation of the mechanism of heat generation in the sun, for which he got the Nobel Prize and which is one of the most important discoveries of the 20th century. Besides that, as a student I had studied some review papers by Bethe that were published in the thirties, which had impressed me by their clarity and rigor, and since then I have always associated Bethe's personality with these attributes. Later I met Bethe several times, mostly in Los Alamos. Bethe had been, almost until his passing away, a regular visitor of the Los Alamos laboratory and worked there on problems of astrophysics. His special links with the laboratory go back to the 1940s when he was the first director of the Los Alamos theory division in the framework of the Manhattan Project. Peter Carruthers, the director of that division during the seventies and eighties, had been his PhD student at Cornell. He was thus after thirty years Bethe's successor at the Los Alamos Lab. I, myself, collaborated in Los Alamos for about twenty years with Peter Carruthers and later also with Dan Strottman and Bernd Schlei my former PhD student from Marburg. Our work had of course nothing to do with the classified research going on there in other divisions.

I also met Bethe once in Bonn in 1984 but this ten minutes accidental meeting has remained engraved in my memory because it threw both of us back fifty years. I had been visiting the Nuclear Physics Institute in Bonn and was waiting for my train to return to Marburg, when I saw on the opposite platform a person I had the impression I knew. Getting nearer I recognized Hans Bethe, who was heading for the Bonn-Cologne airport. We had both just a few minutes before our trains arrived, so we were forced to be brief. Bethe asked me how I, with my Jewish background, was feeling in Germany. I mentioned the Einstein scandal (cf. chapter 18) with which he was familiar and that had surprised him too. Bethe, who was half-Jewish and had to leave Germany after 1933, while his non-Jewish half-brother could stay, told me, with a bitter undertone, that he was on his way back to the US after visiting Tübingen, where the university had just given him back the PhD title, they had taken from him during the Nazi period. (This had happened to all academics of Jewish descent or those political opposed to the Nazi regime).

This unexpected exchange of impressions of two people with similar backgrounds and experiences, albeit of different generations and standing, returned us to a past we would have preferred to forget.

Section VI

Germany's Coping with the Past; The Hydrodynamical Analogy

16

Rewriting History

One of the traumas of post-war Germany has been that of coping with its Nazi past. A simple way of coping with the unpleasant chapters of the past is either to modify them or, more simply, to eliminate them, which amounts to the same thing: rewriting history. This is a well-known practice of dictatorships, which in this way attempt to justify themselves and put in a more favorable light their wrongdoings and crimes. In the Soviet Union, perhaps the first country to apply this practice on a grand scale, the writing of the history of the Communist party is a typical example. Its rewriting was considered so important that Stalin signed the official version in person. One of Stalin's purposes was to portray himself as the closest collaborator of Lenin that was chosen by Lenin to become his heir. His opponents like Trotsky and Bukharin were presented as traitors in the service of capitalism. This practice of rewriting history was applied to all sectors of political and cultural life.

Nazi Germany followed quite closely, although by much more ostentatious means, this practice. The attempts to "arianize" science, the burning of books of Jewish or democrat writers and scientists as well as the exhibition of "degenerate art" of some of the most important painters were nothing else but attempts to deny the role played by these authors and artists in German culture. Wherever this was not possible, because the oeuvre was too well known, other methods were used. Thus Heine's famous poem "Die Loreley" continued to be published in poetry anthologies without mentioning the author's name. It is hard to find a better description of this state of affairs than in Orwell's "1984."

It may sound strange, but democratic regimes are not completely immune of such tendencies, either. Examples in this sense, which I witnessed personally, are:

The denial of the Holocaust. This happened among others both in Germany and in France. It became a topic of hot debate and special names were coined for it: "Auschwitz Lüge" in Germany and "negationisme" in France. In Germany it reached a climax in the famous 1994 Mannheim trial (see below). Things went so far that in Germany, and later in France and other countries, the governments deemed it necessary to forbid by law the perpetration of this denial.

Attempts to attribute the crimes committed by Germany during World War II to the SS only, and to clean the tarnished image of the rest of the army, the Wehrmacht. An exhibition trying to illustrate that the Wehrmacht, too, was involved in these crimes has been the subject of virulent criticism, on the street, as well as in some newspapers. Another illustration of the slowness with which Post-War Germany understood the role of the Wehrmacht is the cultivation of the "tradition" within the *Bundeswehr*, the present army of the Federal Republic and the successor of the Wehrmacht. This tradition has been reflected, among other things, in the names of various barracks of the Bundeswehr: quite a few of these sites have continued to bear the names of generals of the Wehrmacht who have distinguished themselves by their unqualified dedication to Hitler's war efforts, and who often even committed war crimes. Attempts to change these names are meeting even today resistance.[90]

We didn't know. Another aspect of rewriting history is the attempt to deny the knowledge by the German population during the Nazi regime of the existence of camps and of the deportation of Jews, their immediate neighbors, who were deported under their eyes from their homes. In fact, the creation of concentration camps was announced in the German media already in 1933.[91] Heisenberg, personally, had been asked for help by five persons, shortly before their extermination.[92] It is in this spirit of "not knowing" that parents and schools educated for more than twenty years their children, so that this syndrome also contaminated part of the second generation. Fortunately, I have reasons to believe that in the last twenty years the situation has changed. Not only has the tendency of forgetting and ignoring the past been stopped but it has been even reversed. The German grandchildren have started to ask questions and do not accept, anymore, convenient excuses. Teachers and school curricula have changed and the present, third, after-war generation, knows the truth and has learned to live with it, without repressing it. This makes me optimistic about the future of Germany, which I, like most Germans, see within a united Europe.

Hushing up the Nazi past. In a recent edition of the *Deutsche Biographische Enzyklopädie* (K.G. Saur Verlag München, 2001) in ten volumes, the active Nazi past of many prominent scientists of Post-war Germany is completely concealed.[93] To this one could add the claims of some prominent contemporary intellectuals that they became party members without their knowledge.[94]

The German A-bomb

The way the German atomic bomb project was presented by the German scientists who were involved in this project is another illustration of the attempt to rewrite history. Initially Carl Friedrich von Weizsäcker and Heisenberg, in interviews given in the 1950s, among others to Jungk, and which were then reported in Jungk's book "Brighter than thousand suns," tried to suggest that they did not want to build the bomb for Hitler and that they sabotaged the project. This thesis is at odds with various facts, which became subsequently known, and which determined Jungk to accuse Weizsäcker and Heisenberg of a deliberate attempt to mislead him.[95]

In reality,[96] in 1941 Weizsäcker and Heisenberg, like the majority of Germans, were convinced that Hitler would win the war; Weizsäcker had been in favor of the building of the bomb ("At that time I would have considered it, with all determination, an act of cowardice no to do it"[97]) and he had at that time the intention to discuss with Hitler the construction of the bomb and the consequences emerging from it.

Among the facts that contradict the thesis of the sabotage of the atomic bomb project one could quote: This thesis was firmly denied and denoted as "absurd" by other participants in the project, e.g. Gerlach, Harteck and Bagge.[98] In the 1992 interview[97] Weizsäcker mentions a report written by him in 1941 to the weapons office of the army, *on his own initiative*, in which he pointed out that in an atomic reactor one could expect the creation of what is called at present plutonium, and that this atomic element would be particularly useful in the production of an atomic explosive. In the same interview Weizsäcker himself termed his initial claim of the 1950s as "exaggerated." Other significant subsequent statements of Weizsäcker in this interview are: "I took the risk that (out of our work) a bomb would emerge ... Hitler would have had less scruples to use the bomb than Truman. He would have used it without any hesitation. At that time I did not see it in this way." Hitler was for him a "tool of the providence" for achieving power through a political career. Although moral scruples were not the reason why Nazi-Germany did not build the atomic bomb, that did not prevent some German scientists to present it in this way; as a proof of moral superiority, when compared with the deeds of the Americans and English. In the words of Carl Friedrich von Weizsäcker: "History itself will record ... that the peaceful utilization of the Uranium machine took place in Germany (under the Hitler regime), while the Americans and English developed this terrible weapon."[99] Some German physicists and historians went even so far to accuse Einstein of war crimes (cf. below), because he signed the famous letter to Roosevelt, in which he drew the attention of the American president to the technical possibility of developing an atomic bomb and to the fact that Nazi-Germany might work on such a project. These moralists conveniently ignore what could have happened if Germany had indeed succeeded in this project before the allies.

The question of moral scruples has preoccupied not only historians but the public at large as well. It found its reflection among other things in a successful play "Copenhagen" by Michael Frayn, dealing with the meeting between Bohr and Heisenberg in Copenhagen in 1941. It is in part this play, which determined Bohr's family to publish Niels Bohr's memoirs in 2002 already, that is much earlier than the author had disposed in his last will. Among these memoirs is a letter Bohr wrote to Heisenberg in 1954, shortly after the publication of Jungk's book, but which was never posted. In this letter[100] Bohr strongly criticizes Heisenberg's version of their 1941 meeting and in particular the claim that German scientists did not want to construct the atomic bomb.

If moral scruples can be safely eliminated as a possible reason for the fact that the Germans did not build the bomb, what are the true reasons for this fortunate historical fact? Several explanations were presented: 1) the German scientists were not capable to do it 2) the discouraging Nazi environment 3) Hitler's lack of understanding of modern physics 4) lack of resources. A fifth thesis presented by Hermann Jensen[101] is: The danger of a German atomic bomb was much greater than assumed so far. In 1941 Weizsäcker and Heisenberg intended to inform Hitler personally about the importance of the atomic bomb. However, they could not realize this intention because Heisenberg had visited in September 1941 Bohr and had found out he stood firmly on the side of the allies, despite the German victories. Since Heisenberg and Weizsäcker were afraid that they could be accused of high treason because of the unsuccessful conversation with Bohr, during which Heisenberg had handed over to Bohr a design of an atomic reactor, they decided to abandon this project, for the sake of their own survival.

I mentioned Frayn's play Copenhagen also for another, more autobiographical reason: I saw this play in 2003 in Rochester, N.Y. and with this occasion I met Eva Schloss, the step-sister of Anne Franck, who during the intermission was presenting in the theater hall her memoirs "Eva's Story." Eva Schloss, like Anne Franck, had been deported from Holland to Auschwitz together with her parents (the families of Eva and Anne knew each other). While Anne Franck and her mother perished there, Eva (of the same age as Anne) and her mother survived. And what coincidence: after liberation from the camp, Eva and some of her campmates on their way back to Holland passed through Czernowitz. She mentions in her book that the Jewish population of Czernowitz, who had also barely survived, had very well received them. I remember quite well that I have been deeply impressed seeing in 1945 on the streets of Czernowitz the former Auschwitz deportees, in their detainee clothes. This ultimate German phase of the holocaust I had been spared off.

Attempts to Justify the Past

If facts cannot be ignored then one is sometimes tempted to justify them or to set them off against the wrongdoings committed by the enemy. To this category belong:

(1) Attempts to present the bombardment of German cities by the allies during World War II as war crimes, without recognizing that this form of war had been invented by the German air force. The German Luftwaffe started its bombing campaign of civilian targets already on the first day of the war against Poland, on September 1, 1939, with the bombardment and complete destruction of the city of Wielun. This happened even before the bombing of Coventry.[102]

(2) Attempts to present the Germans living in Czechoslovakia and Poland, who were forced to leave their homes at the end of World War II, only as victims, without recognizing their role in the occupation of these countries by the Nazis.

(3) Attempts to justify Hitler's takeover by the hard conditions imposed on Germany by the Versailles peace treaty following World War I. That this war, like World War II, had also been

started by Germany is usually overlooked in these justifications. Among those who put forward this argument, one often finds people like the writer Martin Walser, who are unhappy about the way *official* Germany is coping with its Nazi past. A most recent and quite strident example in this category is represented by the Hohmann scandal of autumn 2003, which produced a vivid reaction (and perhaps over-reaction) from the leaders of the Jewish community of Germany, proving how strongly the past still influences the present. Hohmann, a member of the German Parliament, tried to exempt the Germans of the "guilt" of "people of perpetrators" by stating that the Jews were perpetrators as well: according to Hohmann Jews played a prominent role in the Bolshevik movement and in the execution platoons of the Cheka (the precursor of the KGB). Hohmann ignored the fact that no serious historian has ever claimed that the Germans, as a people, had been perpetrators — perpetrators — by definition, being only individuals.[a] Moreover, as pointed out by the German journalist Henryk Broder, the fact that Stalin was Georgian and had promoted many of his relatives and Georgian friends into leading political positions, would have never induced somebody to claim that the Georgians were a people of perpetrators. The fact that Hohmann used such an argument against Jews was, not surprisingly, qualified as an anti-Semitic act and his political party was forced to expel him.

Ignoring History

Attempts to rewrite history are one way of coping with the past. Like any falsification it is unethical and it ends up by being proven

[a]Hohmann's statement was in part a reaction to the book "Hitler's willing executioners" by Goldhagen, which suggested that Germans were almost by birth anti-Semites and that they carried out without hesitations and scruples Hitler's anti-Semitic crimes. The scientific community of historians had almost unanimously rejected this thesis already several years before.

to be wrong. Still, one could perhaps credit its authors with the intention of *appearing* not guilty of wrongdoings; in an optimistic view this would possibly imply that these people at present know what moral guilt means and that in future they would behave accordingly. That cannot be said about other forms of handling the past. One of these is the fact that the post-war German justice apparatus was not denazified[b]; this lead to long delays in bringing to justice war criminals and, frequently, to very mild sentences. More recent examples are the trials against racist and xenophobic manifestations and against those who denied the holocaust. The Mannheim trial of 1994 brought this state of affairs to a climax, when two judges not only did not send to jail a recidivist perpetrator, who happened to be also a teacher and who denied the existence of gas chambers, but characterized him as a "personality of strong character and high responsibility, with clear principles," for whom political convictions were a "matter of heart." According to the judges, the indicted was "not an anti-Semite in the sense of the national-socialist race ideology but he bitterly held against the Jews their continuous insistence on the holocaust and their financial, political and moral claims towards Germany, which they still derived from it after almost fifty years."[103]

Even at present the German justice has sometimes difficulties in handling such obvious cases as war crimes committed by the SS. In 1944 the SS assembled the citizens of the small Italian village of

[b]Post-communist regimes in Central and Eastern Europe have gone through a development similar, in a certain sense, to that of Post-war Germany. Unlike in democratic regimes, in Russia today the justice system, for example, is yet far from being a third, independent power, besides the legislative and the executive. At present this justice system is still under the influence of the government and any attempt of challenging the presidential regime is annihilated by the organs of the "law." Moreover, in Russia and in the former satellite countries, with the exception of East Germany, the secret services, which were the tools used by the communist regimes to exercise their dictatorships, have not been disbanded but by name and the members of these services have never been punished for the crimes committed. One might argue that the post-war development of the Federal Republic of Germany proves that democratization is a lengthy process and if it took in Germany, after only 12 years of Nazism, decades, it is not surprising that in Russia, after more than 70 years of Communism, it is not yet accomplished. Moreover, even before the revolution Russia (this applies a fortiori also to the other former republics of the USSR) has never been a democracy.

Sant'Anna di Stazzema on the church square, threw hand-grenades into the crowd and then opened fire. 560 civilians, among them 116 children, were killed. Although condemned in absence, in June 2005, by an Italian tribunal as a war criminal, Alfred Concina, a member of the SS regiment, who committed that crime, is still alive and free in Germany. He cannot be extradited to Italy because the German constitution does not permit the extradition of German citizens. Despite the fact that the German justice also started, in 2002, a legal procedure against him, even in 2005 a spokeswoman of the Stuttgart prosecutor in charge with his dossier said that it is not yet clear at all whether this is a case of murder (otherwise it may have come under the statute of limitation) since to qualify as murder the defendant's "unmerciful beliefs" must be proven. The German weekly *Die Zeit* entitled this story[104] quite appropriately "Is massacre murder?"

Another disturbing manner of handling the past is the tolerance of former members of the Nazi party in the highest positions; the fact that Kiesinger had been a registered member of the national-socialistic party did not prevent him from becoming in 1966 chancellor of the Federal Republic of Germany. In the early fifties two-thirds of the high ranking employees of the Ministry of Foreign Affairs of the Bundesrepublik were former members of the Nazi party; some of them were even former members of the SS and SA organizations and condemned war criminals.[105] The history of this Ministry during the Nazi period has become only very recently (2005!) a subject of preoccupations for the Government, when the practice of official "obituaries of merit," which used to be applied also to former Nazis, provoked protests.

That present Germany is not completely immune to this kind of sins of the past is proven by the recurrences of attacks on foreigners, synagogues and anti-Nazi memorials. The extent of extremist incidents, which take place at present, in particular in the regions of the former German Democratic Republic, has often been played down, in part for political reasons, in part out of convenience and sometimes out of sympathy. At a first look it appears surprising that although the Nazi-regime was quite short-lived and found an abrupt end in 1945, its ideological heritage has subsisted so long. The fact that these manifestations of the past are concentrated mainly in East Germany proves that the process of anti-Nazi re-education has been more successful in West Germany. (This is due to the fact that East Germany, the "German Democratic Republic," considered itself *by definition* the new, democratic half of Germany, while the Federal Republic of Germany was according to communist ideology the imperialist half and the successor of Nazi-Germany). If another demonstration of the superiority of democracy as a political system was needed, it has been clearly given by this education success. The two new post-war generations of Western Germany, raised in a democratic spirit, seem to make quite improbable a repetition of history. Moreover, one observes among the young a strong increase of interest in the past: they are asking more and more often their grandparents questions like "why" and "how" the past occured. However, not everybody particularly among the older is happy with this development. Certain intellectuals and politicians are openly "fed up" with being reminded about the Nazi crimes. A typical representative is the writer Martin Walser mentioned above, who condemns the use of the holocaust as a "club" against the German people. People like Walser cannot forgive the Jews Auschwitz.

Misunderstanding the Past

More than any preceding war, World War II produced immense losses of human life and extreme hardships, not only on the side of the Allies, but — albeit on a smaller scale — also among the powers of the Axis. Moreover, like any war, this war too changed the geopolitical structure of the countries involved. Germany, for example, lost part of its Eastern provinces and, more importantly, was divided into two. While it is natural that each side sees and feels primarily its own sufferings, it must not be forgotten or overlooked that this war, more than any other war, had a well-defined "authorship." Unfortunately, until very recently in Germany there had been a strong tendency to look for the culprits in the wrong side.

The hardships suffered by the Germans and Japanese (and to lesser extent by the Italians) were not due to the "others," meaning the Allies, the victors of World War II, but were caused by their own criminal governments, which started the war, confirming the Bible prophecy: "They have sown the wind, and they shall reap the whirlwind." Hadn't been this war, there wouldn't have been the hundreds of thousands of victims of allied bombardments, the displacement of populations from East to West and, last but not least, the Berlin wall and the division of Germany. The same applies to Hiroshima and Nagasaki. This misunderstanding of the past is reflected also in the resentments against the former allies. While initially these resentments were more or less evenly distributed among the allies, soon they concentrated on the Soviet Union. That is due to the role the USSR played in establishing and maintaining the communist regime in East Germany. However, it took a long time until the public opinion of Germany realized that the true culprit of this historical fact was the Nazi-regime itself, which had started the war and was thus the prime responsible for its consequences. It took a very long time until the German people understood what the Second World War was really about. I was personally confronted with this fact, sometime in the early 1980s in Marburg, at an after-colloquium gathering of the Physics department in a *Weinstube*, where to the faculty used to invite the colloquium speaker. I sat next to a colleague, whom I knew quite well since his field partially overlapped with mine. After having exhausted the subject of the colloquium my neighbor said: "Tomorrow we have a mourning day." It sounded like a matter of fact statement, as if one would say the weather is awful. I, initially, did not understand what he meant, but he reminded me that next day was May, 8th. And only then I realized that what for most people meant liberation from the Nazi regime and thus a day of joy, for him this was a day of grief. I tried to make clear to him my point but I am not sure whether I succeeded. That he was no exception is proven by the fact that the President of the Federal republic, Richard von Weizsäcker, the brother of the physicist Carl Friedrich von Weizsäcker, deemed it necessary and appropriate, as late as 1985, to remind his fellow citizens, in a speech which made history, that on May 8, 1945 they were liberated from the Nazi dictatorship and that this date should be remembered as the day of liberation. And even ten years later, in May 1995 I found in "Die Zeit" a mourning announcement with the classical Latin heading "Vae Victis,"[106] in which the 8th of May, 1945 is presented as a memory day "for the fellow-countrymen, who, as a consequence of the 'liberation' (in quotation marks!), were liberated of their belongings, of their homeland and in many cases even of their lives."

I had to think many times about these two ways of interpreting the past. The mourning interpretation reflected of course the fact that during the last days of the war almost two million Germans were displaced from Eastern and Central Europe and tens of thousands lost their lives. (The fact that these sufferings were incommensurable with those, which Nazi-Germany inflicted deliberately upon is not mentioned). But it reflected also a sentiment of injured proud, a heritage of the nationalistic and military ideology of the past, with its misunderstanding of justice, patriotism and honor.

Coping with the Communist Past of East Germany

After reunification in 1989, Germany was faced with another challenge: the bringing into the open of the role played by the Stasi, the security service of the communist German Democratic Republic. It seems that, through an extensive system of informers, the Stasi has permeated almost completely the East German society. It is too early to judge how the reunited Germany has been handling the still continuing, process of uncovering these deeds. The fact that a special state office has been created to make available the documents of the Stasi reflects the importance given by the Bundesrepublik to this chapter of German history and distinguishes Germany from all other former communist countries. What can be said already is that in a certain sense history seems to repeat itself. After 1945 many of the active Nazis and even war criminals denied their involvement in the Nazi regime. So did the tens of thousands of "unofficial collaborators" of the East-German secret police, the Stasi, who denied their informer role in communist East Germany. However, until the contrary is proven, they enjoy the benefit of doubt. This, among other things, distinguishes the democratic society of present Germany from that of the German Democratic Republic. It is a sad irony of history that Nazi-Germany did not need such an extensive informants system. Since the vast majority of the population accepted and was apparently happy with the regime, many also volunteered in reporting to the Gestapo the very few acts of resistance.

17

From Superfluids to Fluids; The Hydrodynamical Analogy Applied to Multiparticle Production in Strong Interactions

---

Learning from the past is not only rewarding in history but in science as well. Since strong interactions have been discovered after electromagnetic interactions the first attempts to treat theoretically strong interactions were based on analogies with electromagnetism. Moreover, one of the most characteristic and important phenomena of strong interactions, multiple production in high-energy particle and nuclear reactions, was *predicted* by Heisenberg in the 1930s, long before its experimental discovery in cosmic rays, by considering the analogy between bremsstrahlung of photons and that of mesons.[107]

Like any analogy, the bremsstrahlung analogy has its limits: it becomes inapplicable when we look at certain details, e.g. the form of the spectrum of produced particles. Actually, Heisenberg himself was quite aware of the limitations of his approach, given the fact that the produced mesons interact while the photons do not, and this lead him to another fruitful analogy, this time between classical hydrodynamics and multiple particle production,[108] since fluids, as distinguished from gases, are strongly interacting systems. Heisenberg drew in particular an analogy between turbulence and non-linear classical field theory. Although the further development of the theory of multiple production did not follow this path,[109] the idea of using hydrodynamics turned out to be very useful. We see here again the "Münchhausen effect": what matters in science is the end result and not the starting point. For the historian of science, however, the "why" and "how" are most interesting.

High energy multiple production inspired also Fermi[110] to draw an analogy, this time with statistical physics: Fermi attempted to treat the system of produced particles as an ideal gas in statistical equilibrium, associating to it a temperature, which depends on the incoming energy. He derived in this way a relation between the multiplicity and energy, which was not very far from the experimental one. However, this approach too was very soon abandoned, among other things because it lead to a wrong particle composition and did not explain why the transverse momentum distribution of secondaries was essentially independent of the center of mass energy of the reaction. Instead of assuming like Fermi global equilibrium, Landau[111] postulated only *local* equilibrium and returned to Heisenberg's analogy with fluid dynamics, leaving, however, completely aside the idea of turbulence, a classical physics phenomenon, which is not well understood even today. Landau solved the equations of relativistic hydrodynamics with initial and boundary conditions appropriate for multiple production and was able to eliminate the most striking contradictions between theory and experiment mentioned above. Indeed, following an argument put forward shortly before by Pomeranchuk, Landau postulated that the fireball, created in the collision process and concentrated in the Lorentz contracted volume, cannot emit particles before it has expanded to a size corresponding to the range of strong interactions, because the particles emitted in the small volume are reabsorbed and do not leave the system. The range of strong interactions is determined by the radius of the pion, the lightest and actually the only meson known at that time. During the expansion the initial temperature cools down from its initial value, which in Landau's formulation is in principle unlimited from above, to a temperature, characteristic for strong interactions, and which is of the order of the pion mass, *the* characteristic mass of the problem. At this point the system starts to emit and the mass spectrum of emitted particles reflects this low equilibrium temperature, i.e. most emitted particles are pions, while the production of heavier resonances, which were later discovered, is exponentially inhibited by the Boltzmann factor. The transverse momentum spectrum of secondaries also reflects the emission from a state of equilibrium: it corresponds to isotropic emission in the center of mass system and is thus again characterized by the Boltzmann factor. This explains why large transverse momenta are also exponentially suppressed and why the average transverse momentum of particles produced is of the order of 300 MeV, i.e. twice the value of the pion mass. It is quite remarkable that more than fifty years after its formulation, Landau's theory has essentially survived, despite the many new discoveries in particle physics.

My interest in hydrodynamics applied to hadronic reactions stems from my work on hot spots in particle reactions. To relate the asymmetry predicted within the hot spot model to the transport properties of hadronic matter, I had used the classical equation of heat transfer, which is a particular case of the equations of hydrodynamics. I was, of course, aware of Landau's hydrodyamical model, among other things because of its relation with statistical approaches to strong interactions. However, the hot spot effect I had been working on (as well as the superfluidity analogy studied earlier) referred to peripheral particle reactions, while Landau's model applied to central collisions. Still, in Landau's 1953 paper there is a remark about peripheral reactions in which the author thanked Evghenii Feinberg for discussions on this subject. Moreover Feinberg had just published a review on "Multiple production of hadrons at cosmic ray energies"[112] and this determined me to discuss with him the hot spot effect. As mentioned in Chapter 15, the correspondence with Feinberg was one factor, which may have contributed to my interest in the hydrodynamical model. Another factor, which I am sure that it influenced my decision to turn to hydrodynamics and the Landau model, was the emergence of new experimental results on multiple production in proton–nucleus reactions. (I had followed these developments also because Erwin Friedlander, my old friend from Bucharest, was among the pioneers of this research). The Landau model was at that time in the process of "resurrection," due to the efforts of Carruthers and Minh, who had shown that the new experimental results on high-energy proton–proton reactions were in agreement with it. It was therefore natural to reconsider this model also in the context of proton–nucleus reactions. For me, personally, this meant in a certain sense a return to nuclear physics, which I had abandoned fifteen years ago, among other things for political reasons. Actually it was still high-energy physics, albeit in nuclei. I was certainly not the only high-energy physicist who had turned his attention to nuclear physics: Nuclei had been, for some time, already, a testing ground for particle physics theories, because they offered the possibility to enhance certain effects observed or predicted in nucleon-nucleon reactions. To quote from one of the pioneer papers by Erwin Friedlander on this subject: "The high nucleon density of nuclear matter compresses the time scale for 'free flight' of high-energy hadrons between their birth and their interactions down to the order of characteristic mean durations of strong interactions."[113] From a particle physicist's point of view, hadn't nuclei existed, they should have been invented.

Classical hydrodynamics describes the motion of fluids. Fluids — liquids, gases and plasmas — are a continuous medium characterized by a specific equation of state, i.e. a relation between thermodynamical quantities like pressure, temperature and density. These quantities depend on the space-time coordinates of an elementary fluid element and therefore hydrodynamics implies only local and not global thermodynamical equilibrium. Often, in particular in the applications of hydrodynamics to nuclear and particle physics, the equation of state is expressed, at fixed number density, in terms of the pressure and energy density. Since for an ideal gas the ratio between pressure and energy density coincides with the square of the velocity of sound, for non-ideal systems, too, the equation of state is parameterized in terms of the velocity of sound. In this more general case the velocity of sound is not constant, anymore, but a function of energy density.

The special role the velocity of sound plays in hydrodynamics is due to the fact that if the velocity of an object in the fluid exceeds that of sound shock waves develop.[114] This phenomenon, well known in hydrodynamics (and in particular in aerodynamics) has been seen also in nuclear matter and confirms the usefulness of the hydrodynamical analogy.

The equations of hydrodynamics are nothing else but the equations of conservation of energy-momentum of a fluid element and this explains why their validity is quite general. Here from also follows that using hydrodynamics for hadronic or nuclear matter is in some sense more than an analogy, it is almost a truism. What is less obvious is the assumption of local equilibrium: One may object that given the short time duration of particle collisions the system has not enough time to reach local equilibrium. This was the main criticism of the hydrodynamical approach, but it was answered by arguing that *strong* interactions make such a rapid equilibration possible. With the advent of the quark-gluon plasma concept the perspective somewhat changed and the justification of local equilibrium became almost obvious: Instead of dealing with four degrees of freedom as in the nucleon-nucleon case (proton, neutron, spin up and down) one had now, depending on energy, between 40 and 88 degrees. This proliferation of degrees of freedom facilitates the establishment of local equilibrium.

The Landau Model Rules the Waves in Nuclei as Well[115]

In a collision between two hadrons the available kinetic energy is released in a Lorentz contracted volume. The mini-fireball created expands according to the equations of classical hydrodynamics and the equation of state of the medium. During this expansion the system cools down and only after a certain minimum temperature is reached particles become decoupled from the system (they "freeze-out") and appear as free particles. In hadron-nucleus collisions the incoming hadron interacts with a coherent one-dimensional tube.[116] This explains why in the nucleus essentially no cascading takes place and why in nucleon-nucleus reactions the multiplicity of produced secondaries exceeds only slightly that observed in nucleon-nucleon collisions.

Landau's original formulation underwent certain modifications: It was found out that only a fraction of the available energy is converted into secondaries, while the rest is taken by a leading particle. This fraction characterizes the *inelasticity* of the reaction and will be commented upon later. Furthermore, Landau assumed that the equation of state of hadronic matter is that of an ideal gas. Expressed in terms of the velocity of sound u this meant that u was assumed to take the value $1/\sqrt{3}$. Subsequently this assumption was also dropped, among other things because of a theoretical argument due to Zhirov and Shuryak who showed that a gas of hadronic resonances lead to a much smaller value[117] of u.

In the proton-proton problem Landau considered the system after shock waves have disappeared. For asymmetric situations like proton-heavy nucleus collisions, this is not possible because the shock waves interfere with rarefaction waves. This interference depends on the size of the nuclear target, giving rise to certain characteristic, size dependent phenomena.

In the mid-1970s, when I started to be preoccupied by hydrodynamics applied to strong interactions, no solution of the equations of relativistic hydrodynamics with arbitrary velocity of sound *and* arbitrary nuclear size was available. Previous work with arbitrary speed of sound had been limited to light nuclei, because in that case the mathematics was much simpler. On the other hand, for applications heavy nuclei were more important, among other things because the probability of producing a system, which shows thermodynamical behavior, increases with the size of the system.

This turned out to be a fortiori true for the search for quark matter with heavy ions, which started a few years later. That is why when I joined in 1974 the University of Marburg I started to think about solving the equations of hydrodynamics for this more general case. I wrote a proposal to the Deutsche Forschungsgemeinschaft and got money for a post-doctoral fellow, who turned out to be, at the recommendation of Geoffrey Chew, Naohiko Masuda, who, after a stay of many years in the USA, wanted to return to Japan. Waiting for an adequate position there he decided to make a stopover in Germany. Although Naohiko had never worked before on hydrodynamics, he soon became an expert and did most of the hard work. That hydrodynamical calculations are not trivial can be seen from the simple fact that the equations are non-linear. Still, for a constant value of the velocity of sound and assuming, as Landau did, that the nuclear target can (initially) be treated as a one-dimensional tube, we were able to find an analytic solution of the equations for arbitrary but constant velocities of sound and arbitrary nuclear sizes, which we used to calculate physical observables. Among the new results we obtained I quote: comparing the number of secondaries in proton-nucleus reactions with those in proton-proton reactions we found that most of the excess in proton-nucleus collisions was concentrated in the backward rapidity region, but that there existed also an increase in the forward region. This result contradicted certain non-hydrodynamical models on the market like the energy-flux-cascade model and the naïve "coherent tube model." Our results also showed that the excess energies were contained exclusively in the backward region. In comparing our results with experimental data we concluded that the Landau hydrodynamical model with a velocity of sound of the order of $1/\sqrt{7.5}$ could explain all the available data.

We also predicted certain new phenomena like: (1) The existence of an asymmetry of average energies of secondary particles between forward and backward regions. Energetic particles should be produced predominantly in the backward region. (2) The existence of an asymmetry in the transverse masses between the forward and backward regions, the sign of which changed, when the speed of sound changed from $u = 1/\sqrt{7.5}$ to $u = 1/\sqrt{3}$. This effect made possible a direct *experimental* determination of the speed of sound in hadronic matter. (3) The disappearance of the leading particle effect in the forward region for small velocities of sound (of the order of $1/\sqrt{7.5}$). None of these predictions has been tested yet.

The above results were published in two consecutive papers.[118] Looking now back to this work it seems to me that one of the lessons I personally drew from it is the realization of the importance of the equation of state. This issue will be addressed below.

Equation of State (EOS) and the Speed of Sound in Hadronic Matter

The purpose of applications of hydrodynamics in nuclear and particle physics is two-fold: explaining and predicting physical observables, and getting information about the equation of state. Unfortunately, the derivation of the equation of state, or equivalently that of the speed of sound, from data is not a straightforward exercise, both for experimental and theoretical reasons. Before the advent of dedicated experiments performed with dedicated accelerators and dedicated detectors a major difficulty consisted in the scarcity of good, reliable data. It quite often happened that, for technical reasons, not all relevant physical observables were measured in one and the same experiment: sometimes, for example, only the rapidity distribution and sometimes only the transverse momentum distribution were reported, not to mention correlation measurements, which were performed quite rarely and with limited statistics.

From the theoretical point of view, besides the big unknown represented by the EOS, the hydrodynamical approach has to cope with the not well-known initial and boundary conditions[119] and the mathematical complexity of the problem, the system of equations being non-linear. For this reason, when calculating physical observables, in the beginning one assumed a constant velocity of sound and one used for the first stage of the reaction a one-dimensional approximation of the equations. In the 1970s and 1980s the validity of these approximations has been challenged and their limitations have been investigated. It became clear that before useful information about the equation of state could be derived from the applications of hydrodynamics, the interplay between these two assumptions — one-dimensionality and constant speed of sound — had to be disentangled. This disentanglement initiated and performed by the Marburg group, proceeded in several steps and took about ten years.

The first step described above consisted in solving the equations of hydrodynamics for arbitrary (constant) values of the velocity of sound and arbitrary atomic mass numbers for proton–nucleus reactions in one dimension. This proved the great sensitivity of certain physical observables to the value of the speed of sound. A second step, to be presented below, was the inclusion of a phase transition in the equation of state, leading to a change of the velocity of sound during the expansion process.

The equation of state determines, among other things, the form of the energy distribution of secondaries and the value of the initial temperature of the fireball. The use of the energy distribution for the determination of the equation of state (of hadronic and nuclear matter) was exemplified in the papers with Masuda on proton–nucleus collisions quoted above. Soon after this, new proton–proton data at the CERN Intersecting Storage Ring accelerator ISR became available, which made possible not only the determination of the initial temperature of the fireball at the highest energy available until then (63 GeV center of mass energy), but brought first evidence for a change with temperature of the velocity of sound, and thus of the equation of state.

Landau, not knowing about hadronic resonances, which were discovered several years after he wrote his 1953 paper, assumed that hadronic matter obeys an ideal gas equation of state. In a certain sense, however, this assumption was also quasi-prophetic:

The quark-gluon plasma, the existence of which was predicted two decades after the discovery of resonances, obeys an ideal gas equation of state[120]! Moreover, Heisenberg and Landau made another bold assumption, which at their time was not easy to defend: they postulated that local equilibrium, the pre-condition of the applicability of hydrodynamics is rapidly established. According to our present understanding, this local equilibrium is a consequence of the large number of degrees of freedom, which characterizes the quark-gluon plasma.

Initial Conditions in the Landau Model: Inelasticity and Lorentz Contraction

Initial conditions are a prerequisite of solving any equation of motion including those of hydrodynamics. In high-energy physics the main initial conditions refer to the initial volume and the energy available for particle production. What distinguishes macroscopic hydrodynamics from that applied to particle and nuclear reactions is that in the latter, due to the high energies involved, Lorentz contraction of the initial sizes of the colliding particles plays an important role. As a matter of fact, Landau calculated this contraction using the classical Lorentz-Einstein formula in which the contraction factor is determined by the velocity or the energy of the incident particles. As to the amount of energy available for particle production, he assumed that the colliding particles are fully stopped and therefore their entire incoming energy is available for particle production. With these assumptions he could explain the main features of multiparticle production. In the early 1980s, new data at much higher incoming energy became available and a comparison of the new data with Landau's theory showed some disagreement: the mean multiplicity did not increase with the incoming energy E as fast as predicted by Landau ($\sim E^{1/4}$). Given my previous experience with the hydrodynamical model I could not believe this result. I happened to be in good company with this skepticism of mine: Erwin Friedlander did not believe it either. We observed, first of all, that the experimentalists had not taken into account a fact, which had been discovered in cosmic rays soon after Landau wrote his pioneering paper: The assumption of full stopping in nucleon-nucleon and nucleon-nucleus collisions did not hold. In essentially all events "leading" particles were found, which took about half of the available energy. By reducing the amount of available energy by an inelasticity factor K, representing the ratio between the energy effectively used for particle production and the total energy, we could re-establish in part the agreement between theory and experiment. However, at very high energies we found that even this reduction of available energy due to the inelasticity factor was insufficient to account for the experimental observations. Therefore we conjectured[121] that the mean inelasticity, assumed until then by everybody to be constant (or order ½), was in fact itself a (decreasing) function of energy. This conjecture turned out to be quite fruitful: in the same accelerator experiment where the initial disagreement with Landau's model had been reported, evidence for such a decrease of inelasticity was subsequently found. Encouraged by this development the Marburg group developed in the 1980s and 1990s in several papers,[122] a theory of this new effect and in particular a theory of the inelasticity distribution, in terms of an *interacting gluon model*, based on QCD. This model, known in the literature under its abbreviation IGM, has found followers and is currently still of actuality. This story is another example of the usefulness of the Landau model.[123]

Local Equilibrium — The LESIP Meetings

We have seen that the application of hydrodynamics to the microcosm presupposes local equilibrium. This is also true for the use of statistical concepts in general. The assumption of local equilibrium appeared difficult to be reconciled with the idea that the mean free path in high-energy reactions is comparable with the dimensions of the system and thus the system has, apparently, not enough time to equilibrate. This idea was particularly popular among some nuclear physicists who assumed that they could calculate a mean free path by using the known cross section of hadrons in nuclear matter. That assumption was of course at odds with the fact that cross sections are defined only for particles *on shell,* that is for free particles, but in the heat of the argument this was often forgotten. A related and equally misleading preconception was the belief that for statistical equilibrium heat baths are necessary. Actually even today, when the application of thermodynamical and statistical methods in heavy ion physics has become standard, one still meets from time to time in the literature, although much rarer than in the past, reservations about the use of these methods, based on distinctions between the concepts of "phase space dominance" and "true" thermodynamical behavior. I was from the very beginning convinced that these objections were unjustified for the simple reason that entropy[124] is by definition a statistical concept and from it temperature is derived. Therefore, except for fluctuations, there is no difference between the concepts mentioned above. This presumably explains why the pioneers of the field like Heisenberg, Fermi, Landau, Hagedorn had used statistical and hydrodynamical concepts, without any reticence. However this was not sufficient to convince the "community." From my own experience as a teacher of theoretical physics I believe that this situation is due in great part to the unilateral and overspecialized education of the generation of high-energy and nuclear physicists of the 1960's and 1970's, in particular in the West. The problem became more acute with the opening of a new chapter in strong interaction physics, the search for quark-gluon plasma. One envisaged producing this new state of matter in the lab through high-energy collisions, albeit of heavy ions. While the assumption of local equilibrium appears easier to be accepted for heavy ion reactions than for particle reactions, the proponents of these new experiments had a tough time to get their proposals accepted: To produce phase transitions at an accelerator sounded for many physicists like science fiction. This situation as well as my preceding work on hot spots determined D. K. Scott and me to organize the first LESIP meeting. We had the chance to be able to gather *the* experts in this field, among others Carruthers, Feinberg, Sewell and Shuryak and the success of this meeting encouraged Peter Carruthers and me to make of this type of meetings a series.[125] The second meeting was held in 1986 in Santa Fe, New Mexico, the third one in 1988 in Tucson, Arizona and the last one in 1990 in Marburg.

I quote below from the introduction of the talk by Geoffrey Sewell at LESIP I: "My essential objective, in this talk, is to demonstrate how both local equilibrium and hydrodynamics arise, under rather general conditions, as a consequence of the dynamics of many-particle quantum systems in pure states. These states may, of course, be very complex, but the point I want to make is that the occurrence of local equilibrium is not dependent on the assumptions of heat baths, Gibbs ensembles, etc., that form the standard paraphernalia of statistical mechanics: the interactions between local sub-regions of a system, in appropriate states, possess all the randomness we need. I would imagine that this observation is relevant to the subject of the present meeting, which is concerned with the occurrence of local thermal and hydrodynamical behavior in nuclear and high-energy physics, in situations where there are no heat baths provided."

Feinberg addressed among other things the issue of thermalization in particle physics as compared with thermalization in heavy ion reactions. After mentioning the amazing success of hydrodynamical and thermodynamical descriptions in explaining certain experimental facts like the limitation of transverse momentum of secondaries in particle reactions, for which no other theoretical interpretation was available, he pointed out that perturbative QCD was not an adequate tool for describing the initial stage of a collision, and that multiparton collective interactions had to be considered; these could provide a mechanism for randomization and equilibration. Along similar lines Shuryak invoked the important role of the QCD vacuum, which increased dramatically the number of degrees of freedom of the system and thus made thermalization possible. The summary talk given by Peter Carruthers "Is local equilibrium a useful concept in hadronic interactions?" answered this question in the positive. Three round tables, the titles of which were self-explanatory, contributed significantly to the clarification of unsolved problems. These were: (1) Mean free path in Strong Interactions; (2) The Landau Hydrodynamical Model; (3) Perspectives of Strong Interaction Physics. I expressed the opinion that the discussion about the mean free path in strong interactionsvery much resembles the construction of the Babel tower: people speak different languages and often what they do is not very useful and sometimes even confusing: although in the literature one finds statements that the mean free path is an experimental quantity that can be measured, the truth is that the mean free path is a theoretical concept. For strong interactions in particular, where no theory outside the perturbative regime of QCD exists, the mean free path, which is defined in analogy with the game of billiard balls, is useless and misleading: it is an example of the limitations of certain analogies. Shuryak added to this the remark that instead of using the ill defined concept of mean free path, one should rather use transport coefficients and correlation lengths.

Another topic, which came up at the first LESIP meeting in the discussion of the Landau model, and which will be considered below, was that of Lorentz contraction in the initial stage of the collision. This issue is important because the Lorentz contracted volume of the colliding particles determines the initial energy density reached in a collision. Here, again, it was pointed out that only a (still non-existing) theory of strong interactions could provide a reliable estimate of this effect. Naïve classical considerations could be misleading. Related to this is the question of inelasticity, which determines the amount of energy available for particle production. Here some new results of the Marburg group were discussed. The outcome of the meeting was that although no rigorous derivation of local equilibrium was possible at that stage (this situation has not changed in the last twenty years), there existed general plausible reasons to expect local equilibrium and that the data left little alternative for this assumption. Another lesson was that both particle and nuclear high-energy physics needed that assumption.

The following LESIP meetings continued up to a point the discussions on local equilibrium and the applicability of statistical concepts, but they also reflected the evolution of preoccupations of the community, as illustrated in the titles of the new meetings. One new topic of the LESIP meetings, which emerged quite strongly, was that of correlations and in particular that of Bose-Einstein correlations (BEC).[a] LESIP IV was organized with the occasion of my 60th birthday. I remember this meeting also because of a very funny after dinner talk by Peter Carruthers and the caricature portrait he drew of me I still possess.

The hydrodynamical method, based on an analogy with classical fluid dynamics, has not only proven to be quite fruitful in multiparticle phenomenology, but it lead to predictions of some new effects like solitons, also based on analogies. Moreover it turned out to be an indispensable tool in the search for the new state of matter, quark-gluon plasma, which owes its name to another analogy. These topics are the subject of the following chapters.

[a]The abbreviation BEC is sometimes also used in the literature to denote Bose-Einstein condensates. In this volume we will use it exclusively for Bose-Einstein correlations.

Section VII

Einstein in Marburg; Solitons; The Auschwitz "Experiments"

18

Einstein Criticized in a Marburg Colloquium

It didn't take us a long time to get accustomed to life in Germany and to Marburg in particular. We didn't mind the smallness of the town since our professions absorbed us completely, weekends included. Besides that we, and I in particular, travelled a lot and enjoyed the freedom of movement, which I had missed so much. Frankfurt airport was only one hour from Marburg after all, and that reminded me of Bleuler's remark about the five trains per day from Bonn to Switzerland. However after only five years of Marburg I was caught up by Germany's past. To my surprise I discovered that this past has had a strong influence not only on its after-war political life, but also on its after-war scientific establishment. Moreover this influence did not limit itself to social sciences, but it permeated also natural sciences and medicine. I will discuss two cases, one in physics, in which I had the doubtful privilege to be personally involved, and another in medicine, about which I learned only from the media, but which is much more serious. Both these examples illustrate that even decades after Hitler some German scientists, too, had difficulties in coping with the past of their science.

In spring 1980 a German astrophysicist, A. Unsöld, gave a colloquium at the Physics Department of the University of Marburg on "Evolution of cosmic, biological and mental structures." The somewhat unusual title for a physics colloquium as well as the name of the speaker — Unsöld was a well known, albeit retired, astrophysicist — attracted an audience larger than usual. The speaker presented a model of the evolution of the universe the details of which I do not remember, possibly because the model seemed to me not very interesting or new. I started to pay more attention when Unsöld began to apply his considerations to the evolution of mankind and then to the evolution of the human brain: He ended up in "deriving" from the evolution of the cosmos certain conclusions about the psychology, ethics and behavior of scientists and then strongly criticized the acts and conducts of certain scientists who put their discoveries or inventions at the disposal of governments. Eventually, he concentrated on two examples: Fritz Haber, because he invented the poison gas during World War I and Albert Einstein, because he signed in 1939 the famous letter to Roosevelt in which he attracted the attention of the American President to the fact that Germany is able to develop an atomic bomb. According to Unsöld these acts were not only immoral but also plainly criminal. In his words: "The crimes of Fritz Haber and Albert Einstein were not less serious than those of Adolf Hitler." At this point I could not remain silent anymore. I addressed myself loudly to the chairman of the colloquium, Walcher, asking for a correction. Walcher did not react and the speaker continued his talk. Strangely enough, I was, apparently, the only one in the audience to protest. I had no other choice but to ostentatiously leave the auditorium. This must have been observed by the entire audience, including the chairman, because I left through the front door of the hall. However, when later asked about this incident, Walcher at first denied categorically that this happened but eventually, after being confronted with the statements of several people who witnessed the incident, changed his mind (cf. below).

The fact that I had been the only one to protest was for me an unpleasant surprise and I could not avoid the feeling that most of the audience agreed with the speaker's statements. Still, in the evening of the same day I got a call from Reinhard Brandt, a professor in the Nuclear Chemistry Department, who expressed his solidarity with my stand and who afterwards became very active in the follow-ups of this affair. I later learned that after me several others had left the hall, but whether this was in protest or just because the talk was too long, I don't know. Although rather shocked by this incident, I was initially tempted to consider it a slip of the tongue of an old man, who just wanted to make a rhetorical point. Seven months later I was forced to realize that I had been wrong. Indeed, in the November issue of *Physikalische Blätter*, the journal of the German Physical Society (the equivalent of *Physics Today* in the USA) the same Unsöld published a paper[126] in which he not only repeated his criticism of Haber and Einstein, but made certain statements about Einstein in particular, which not only were unfounded and slanderous, but which reflected a profound sentiment of hate.

Although he acknowledged the great merits of Einstein's scientific work, Unsöld expressed his disappointment that during the 1979 celebration of the hundreth anniversary of Einstein's birthday, almost nobody mentioned that Einstein's name is to a large extent related to the atomic bomb. Referring to Einstein's letter to Roosevelt, Unsöld asked: "Could a physicist with the intelligence of Einstein not have foreseen what would happen in 1945 in Hiroshima and Nagasaki?" When discussing the responsibility of scientists for the application of their discoveries Unsöld claimed that "the author of a development project has from the beginning until the end the total responsibility for his activity, and in this sense one can say that Einstein shares the responsibility for the development of atom bomb. Many would even consider today the expression of just 'alas' a strange reaction, given the 300,000 deaths. Later Einstein apparently repressed (in the Freudian sense) the whole problem."

Analyzing Einstein's scientific activity Unsöld finds his early papers deficient because he did not quote the work of his predecessors. In this Unsöld sees a manifestation of Einstein's "narcissism." According to Unsöld, Einstein, after reaching the age of forty-five, had no new ideas anymore. Because of that he turned his attention to public affairs and in particular to "Jewish nationalism," quite similarly, in Unsöld's view, to what some German physicists like Lenard, Stark and Wien did, when they embraced nationalistic and anti-Semitic ideas. Einstein moved from Prague to Berlin in 1914, not so much because of scientific interests, but because of money. Money was also the reason why Einstein wanted the Nobel Prize. After 1920 Einstein made frequent trips abroad "to fill the emptiness created by the drying up of his more profound ideas," to make "propaganda" for the theory of relativity and for the creation of a Jewish state. Comparing the fate of scientists who had to flee from Nazi Germany with the fate of those who could remain, Unsöld writes: "perhaps it should be appropriate not to show compassion only for the emigrants of that time, a large part of whom, fortunately, found quite soon good positions. It was not always easy for the scientists who remained in Germany to continue to do research, some teaching and, last but not least — besides the work demanded by the Wehrmacht — to secure the often irreplaceable libraries and the scientific material in the bombed cities."

This time I was not so much shocked by what Unsöld wrote — my colloquium experience had left me no doubts about his opinions — but by the fact that such an article was published in the official journal of the German Physical Society, albeit under the heading "For discussion." I started to ask myself whether I was perhaps wrong in my negative reaction and that Unsöld's points were not so outrageous, after all. To convince myself I translated the article in English and sent it to a few colleagues abroad. Their reaction was immediate and unequivocal: Not only was Unsöld wrong in many of his statements about Einstein, but the fact that he had singled out two Jews, Haber and Einstein, to prove the "criminal" behavior of certain scientists, reminded of times one had thought long passed. What amplified my shock was that after the publication of Unsöld's article in the 1980 November issue of *Physikalische Blätter* two more issues of this journal appeared, in December 1980 and January 1981, without any comment or reaction of the editors or readers to Unsöld's paper. Moreover, in a kind of summary of the events of 1980, signed by the president of the German Physical Society who happened to be Rollnik, one finds the following remark: "There has been only little criticism about the selection and the content of the contributions.... "All this determined me to write a letter to Rollnik. Here are extracts from my letter of January 26, 1981: "I have read with amazement the paper by Unsöld in *Physikalische Blätter* of November 1980 and I am even more surprised that so far there hasn't been any reaction to it in this journal. I think that the paper contains untruths and statements, which can be found in a *National Zeitung*[a] or in K-newsapapers[b] but which have no place in an official journal of the German Physical Society. In the discussion of Einstein's letter to Roosevelt the author does not mention its decisive motivation, i.e. the fact that fission of uranium had just been discovered in Germany and that Hitler himself had the possibility to start the construction of the A-bomb and effectively tried to do that. If one talks about crimes, as Mr. Unsöld does, then this conscious disinformation of the young and probably uninformed generation (not surprisingly if it had teachers like Mr. Unsöld) can hardly be characterized differently. That Einstein later suppressed this problem is totally false. The statement that resistance during the Third Reich would have been 'senseless suicide' is a mockery of the admittedly few, but the more so to be admired, martyrs, who gave their lives, saving the German people of a collective guilt. The self-pity of the Unsöld generation is not a good example for the present generation. In statements asserting that Einstein needed the Nobel Prize to divorce his first wife, or that money and not scientific reasons were decisive in his

[a] *National Zeitung* is a German extreme right-wing newspaper.

[b] K-newspapers (abbreviation from "communist," spelled in German "Kommunist") refers to extreme left-wing newspapers.

movement to Berlin, in insinuations that Nazism equals Zionism or that 'the emigrants found soon good positions' it is difficult to decide where the ridiculous and tastelessness end and the hate and malice begin. After reading this article one can hardly avoid the impression of being faced with an intentional and systematic slander of a great personality, whose memory is dear to the entire civilized mankind. For your information: In May[127] 1980 Mr. Unsöld gave in Marburg a colloquium in front of more than one hundred people, among them a large number of students, and said word for word: 'The crimes of Haber and Einstein are not less serious than those of Hitler.' Unfortunately, I was the only one who protested; I had no choice but to ostentatiously leave the hall. Only one colleague called me later up and expressed his solidarity with my stand. I consider the publication of this article in *Physikalische Blätter* a very grave incident, which has nothing in common with a 'discussion.' I take the liberty to give you and the editorial and publishing boards the friendly advice to distance yourself strongly, explicitly and in public from this article and to take steps that such regrettable incidents do not happen again. Otherwise the prestige of physics in Germany will suffer further damage. For me and others who think like me, the continuation of membership in the German Physical Society would not be possible anymore. I expect you to forward this letter to the members of those boards."

This time Rollnik's reaction was swift, but for me even more perturbing. Only 24 hours after I had posted my letter to him, he called me up to tell me that the publication of the paper had occurred because of a technical mal-functioning in the work of the editorial board, that in the article there were certain "subtleties," which could be perhaps interpreted in a manner not intended by the author. That according to Paul, who knew Unsöld from the pre-war time, Unsöld was not an anti-Semite; that Unsöld had been fed up with the avalanche of manifestations to celebrate Einstein's hundreth anniversary. The real surprise in this telephone conversation emerged when I asked Rollnik why the boards had not reacted so far. The answer was: "You happen to disagree with the content of the article, but you are an exception. I got dozens of letters expressing support for Unsöld's point of view."

One or two days after Rollnik's call I got a letter (dated January 29) from W. Hanle, the member of the publishing board of *Physikalische Blätter* responsible for history of science, stating that in hindsight, he regretted the publication of Unsöld's article. Hanle had had a long exchange of letters with Unsöld, trying to induce him to make changes in his article, but Unsöld, essentially, refused. The publication under the heading "For discussion" was a compromise. According to Hanle, "In its original version the article was even less acceptable." Hanle himself was in a "particular situation," because at the age of 21, in 1922, after a talk about the theory of relativity he had given in Heidelberg, he had a strong conflict with Lenard and was forced to leave the university. My letter to the president of the Physical Society (and a similar letter by Peter von Brentano, a colleague from Köln, cf. below) was a "justification" for his opposition to publish Unsöld's article. Such a justification seemed necessary, as there were also "enthusiastic" letters of support for Unsöld.[128]

Brenatno's letter to Rollnik about which I learned from Hanle (as far as I remember, Rollnik had not mentioned it to me) had been written already on November 17, 1980. It stated among other things that the article by Unsöld consists of a chain of attacks against the personality and integrity of Albert Einstein, which are unbearable and unacceptable. To avoid further damage the author asked among other things that the executive board of the German Physical Society should distance themselves from Unsöld's contribution and that a distinguished historian of science should be asked to characterize Einstein objectively. Brentano's letter which intended to be a "letter to the editor" was not published but eventually the editors implemented the proposals contained in it. However my telephone conversation with Rollnik, which took place two months after the letter from Brentano, made me doubt whether this would have happened, had not been another factor, which convinced the editors that they had to act: the reaction from abroad. Once the article by Unsöld became known there had been an avalanche of letters of protest addressed to the president of the Physical Society and to Herwig Schopper, the director general of CERN, who happened to be at that time a member of the publishing board. The authors of these letters were prominent scientists from all over the world. I have copies of some of these letters of protest, which criticize in very strong and unequivocal terms both the article by Unsöld and the fact that it was published. To quote some characterizations of Unsöld's article: "The article is a scurrilous mixture of pseudo-history, attacks on Einstein's personality and personal life, and apologies and false analogies. Einstein is Unsöld's apparent target, but one must be blind, indeed, not to see the larger aim." "This article it appears to me a senseless and tasteless diatribe on one of the most revered members of the profession of physics. My concern is not so much that an author can hold such views, as it is that some seal of respectability can be given them by their acceptance for publication in an official journal of a professional society so prestigious as the German Physical Society."

Some of the authors of these letters made a point in emphasising that they were not Jewish. An article, which Rolf Hagedorn sent me, and which he submitted for publication to Physikalische Blätter, impressed me in particular. This article not only put the record straight, but was also written with much wit. In the accompanying letter to Rollnik dated February 4, 1981 Hagedorn asked that his article be published "not as a reader's letter, but as a leading article." He also asked the editors to publish in the same issue Einstein's letter to Roosevelt. Hagedorn's letter ends with the sentences: "Should it be necessary for me to join the German Physical Society in order to have the article published, please send me the appropriate forms. I hope to be able to get letters of recommendation."

Since Hagedorn's wish to have his article published was not fulfiled by the publishers of *Physikalische Blätter*, I think it appropriate to do it now here.

Hagedorn's Paper Submitted for Publication to *Physikalische Blätter*[129]:

Albrecht Unsöld — what follows after that?

"Albert Einstein — One Year Later" is a mixture of resentments, apology, psychology, history and moralin, selected, scrambled and presented to the reader by Mr. Albrecht Unsöld, an internationally known scientist. What has happened?

Does Mr. Unsöld want to convince us that too many nice things have been said about Einstein during the Einstein year? Or that Einstein was only a compiling genius — quite typically without references — of what others knew before him? Or that at the age of forty five the production of new ideas ceases and the frustrated scientist then writes essays against Einstein — sorry — I mean that he turns to public life? Or that Einstein had — how repugnant — high salary demands at his appointment to Berlin? (If somebody is so fortunate to be Einstein he should not ask also money for that. Mr. Unsöld would have been certainly happy to pay for becoming an Einstein, and then he would even have renounced the Nobel prize, instead of using it, immorally, for his divorce). This Einstein was anyhow a fellow without fatherland, who already in 1915 pursued the division of Germany[130] *and who, once he had no new ideas anymore, made propaganda trips abroad for the international Zionism and for his general theory of relativity.*

Unsöld: "It should be also permitted to ask how the Federal government would treat a 'civil servant' with such views,[131] *be he Nobel laureate or postman? "When Hitler chased Einstein out of Germany" — quite an understandable act (R. H). in view of what was mentioned above — "and Einstein moved to Princeton," — how could it be otherwise — "the divorce between Einstein and Germany was complete." Obviously: such a guy grasps with the first occasion the atomic bomb. Those who had remained in Germany and who continued, "besides fulfiling the requirements of the Wehrmacht," to do quietly and painfully research and a bit of teaching...were nobler. (To avoid any misinterpretation: I also fulfiled as a soldier the requirements of the Wehrmacht,*[132] *I, too, was not in the resistance; then, however, I do not dare to criticize Einstein on moral grounds).*[133]

Enough; all that is not what Mr. Unsöld is after; it serves only to prepare the main coup: he establishes Einstein's co-responsibility for Hiroshima and Nagasaki. One might ask of course: how much did Einstein's signature under the well known letter to Roosevelt contribute to the atomic bomb and how big is Einstein's responsibility? Unsöld is right when, before answering this question, he distinguishes between research and development and considers someone who pursues a certain development towards a given goal (e.g. when he "fulfils the requirements of the Wehrmacht" R.H). *more responsible than somebody who, while doing pure research, makes a discovery, which can be misused. Let us make Einstein accountable! Einstein signed the letter of August 2, 1939, when the war was obviously inevitable. Einstein knew that the fission of uranium made the atomic bomb possible and he could not have doubts that the German physicists — whom he knew — were also aware of that and were capable of constructing the bomb. There were even indications that they had started doing that. Einstein knew Germany, he had lived the beginnings of the Nazi era and he knew enough about its further development to figure out what Hitler could do with the atomic bomb and how the world would look like after his victory. At that moment it was everybody's guess that the only thing that mattered was to get a however small lead.*

For the sake of this lead Einstein signed; hadn't he done that Roosevelt would have acted a few months later — the lobby existed and the bomb would have come anyway. That is how Einstein must have seen it. What would you, Mr. Unsöld, do in such a situation?

One could seriously discuss such problems and even make Einstein co-responsible. I myself, however, wouldn't do that, because under the given circumstances he must have acted as he did; moreover, he did not participate in the further development, which, I repeat, would have taken place also without him; in particular he did not participate in the discussions of those who decided to drop, without warning, the first bomb on Hiroshima, and shortly after that on Nagasaki. Today every second physicist works for the arms industry; that is a topic, Mr. Unsöld, one could write about a lot.

I have nothing against criticism of great men; what I hold against you, Mr. Unsöld, is that you gleefully construct the proof of Einstein's co-responsibility, by destroying him at first morally. In this way the young reader who has not lived during that time should realize whom he is dealing with. Where is your feeling for proportions and orders of magnitude?

At the end of your article you become again concrete, with neocortex and things like that. This last paragraph is very instructive, indeed. One understands how your essay came about: your limbic brain run off.

You may perhaps be right with your statement that at "the present stage of the evolution of mankind" the general question about the moral responsibility of science and of the moral responsibility of politics cannot be settled yet. On the other hand the question of Mr. Unsöld's moral responsibility for his essay has an answer: this essay is morally irresponsible. Why? Unfortunately, we have now again brown shirts, the Hitler salute and groups of Wehrsport,[134] *even bloody attacks on synagogues, parties, and railway stations. I cannot imagine that you agree in the slightest degree with any of these; however, have you considered what effect your essay would produce on those people who do not condemn or who even sympathize with this development? Mr. Unsöld, you say the truth in each of your points, but how you say it and what you omit to say, this will bring you applause from the wrong side. This is also important to realize.*

ALBRECHT UNSÖLD — WHAT FOLLOWS AFTER THAT?

I am asking.

Rolf Hagedorn

The publishers not only refused to publish Hagedorn's paper, but they decided not to publish any papers or letters to the editor on this subject. Instead, they published a "final" statement in the March issue of *Physikalische Blätter*,[135] which says that the Unsöld article had been published because of a "failure of coordination," that it contained wrong, and therefore unacceptable statements and that it should never have been published "under this form." The article did not reflect the opinion of the editors and a historian of science will be asked to analyze the relationship between Einstein and Germany.

A short time after this statement, however, several important foreign science journals took up the issue. In an editorial "German Physics in Row about Einstein" *Nature*[136] reported the events quoting some of Unsöld's statements, including the comparison between Einstein and Hitler, and mentioned that "there was no suggestion that (*Physikalische Blätter*) will publish the critical letters," although there had been "a torrent of protests against the publication of Unsöld's paper." Similar reports and strong criticism appeared in *La Recherche*,[137] *Discover*,[138] *Interdisciplinary Science Reviews*.[139] A few months later the German monthly *Bild der Wissenschaft*[140] also published critical comments by two German physicists.

Although the editors initially refused to publish any critical letters — despite the fact that Unsöld's article had been published "For discussion" — they soon had to change their mind, because of a protest, this time from Unsöld himself, who made use of his legal right to respond to the statement of the publishers. In his response[141] Unsöld writes that he had received a dozen of letters from "very renowned scientists" who expressed their strong support for his article ("only two of them made some critical observations about details"), as well as many similar oral comments. While stating that he had not intended to denigrate the personality of Einstein, Unsöld continues to criticize attempts to create a "hero legend" about Einstein.

This time the editors published (on the same page) a letter by Rudolf Haag and Harry Lehmann, which criticizes Unsöld's paper along the same lines as those mentioned in the letters by Hagedorn, by Brentano and myself.

A similar "exchange" of criticisms took place in *Nature*. Unsöld wrote a letter to the editor[142] complaining about the editorial and essentially repeated his previous assertions. Moreover, he attempted to negate the walkout incident by stating "in the discussion which followed no one from the audience of some 200 said anything about Einstein." On the same page of *Nature* appeared a letter by the managing editor of *Physikalische Blätter* Kromphardt, a former student of Unsöld and whom the president of the German Physical Society considered responsible for the publication of Unsöld's paper. Kromphardt criticizes the president's interference with the publication policy of *Physikalische Blätter*, praises Unsöld's article because it "destroyed the *idol* 'Einstein' in order to get at the main theme... the increasingly dangerous disproportion between man's moral capacities and his intellectual faculties." Although claiming that Unsöld's article "was not intended to be anti-Semitic," at the end of his letter Kromphardt becomes more explicit: "Surely it was also tragic that nuclear weapons were developed by leading scientists of a people Hitler had horribly plagued. Nevertheless for us in Central Europe this is of little help, if these weapons are to be used in our countries. Not even the 30,000 Jews who still, or once again, live in Germany could hope to be spared."

Unsöld's letter to the editor was answered, in *Nature*, in a letter[143] written by some members of the Marburg Physics Department and by Brandt (12 out of a staff of about 100; there was no senior professor among those who signed), in which they distanced themselves from Unsöld's views. In February 1982 *Physikalische Blätter* finally published the paper[144] on Einstein by a historian of science, as demanded by Brentano and as announced in the president's statement of *Physikalische Blätter* 37 (1981) 85. The author, A. Herrmann,[145] points out, among other things, that Unsöld's references on Einstein's biography were mostly "second hand" and that Unsöld himself had apparently not read most of the relevant documents. Herrmann sketches a short history of Einstein's relation with Germany and points out that despite all the evil done to him in Germany, there were mostly German scientists who had brought about his recognition.

Imre Toth, the distinguished historian of science whom I visited in Princeton in 1981 during his sabbatical at the Institute for Advanced Studies, got involved in this story from the very beginning. He was so much revolted by Unsöld's article that he decided to give a course on "Einstein and the Third Reich" at the University of Regensburg, which attracted a very large number of students ("you made me sick" was his repeated remark in some of the five letters he wrote to me on this subject, after I sent him a copy of Unsöld's paper). In his letter of March 27, 1981 Toth mentioned that he had met Helene Dukas, Einstein's "still very vivid" lifelong secretary, (she died one year later at the age of 86) who was "very much consternated" about the Unsöld affair. She showed him a whole file on this topic, containing among other things an article from *Aufbau*, the famous journal edited by German immigrants, which at that time still appeared in the US. I assume that this file is now part of the Einstein Archives, which are deposited at the Hebrew University in Jerusalem. Imre Toth discussed this story also with Abraham Pais at the Princeton Institute and John Stachel, the editor of Einstein's Collected Papers. Both were "disgusted."

The ups and downs in the Unsöld versus Einstein affair continued. On March 28, 1982 the liberal German weekly *Die Zeit* published a short article signed by the science editor of the journal Günter Haaf. The very title of this article "Censorship in a professional journal; is criticism of Einstein a sacrilege?" contains the essence of the message the journalist wanted to convey. Haaf reports that a group of members of the German Physical Society consider the way the Unsöld affair was handled in *Physikalische Blätter* an interference of the executive board of the German Physical Society in the functioning of this journal and that a motion demanding the reestablishment of *Physikalische Blätter* as an "independent journal" will be presented at the next meeting of the German Physical

Society, In Haaf's words: "There is censorship in *Physikalische Blätter.*"

After quoting — without any attempt of qualification — Unsöld's attacks on Einstein, Haaf characterizes Rollnik's actions after the publication of the Unsöld article, in particular the statement that this article should never have been published, as censorship. That Unsöld's article was full of resentments, that it had stirred strong protests of leading scientists in and outside Germany, that it had been vividly criticized in authoritative foreign editorials, is not mentioned in Haaf's article with a single word.

The article of *Die Zeit* was a double disappointment for me: firstly it confirmed something I had until then refused to believe, despite Rollnik's telephone call and the repeated warnings of Brandt: That Unsöld was not an isolated case among German physicists and that he had many followers. Besides that I started to wonder about the liberal attitude of "*Die Zeit.*" As to this last aspect, I still hoped that Haaf's article did not represent the opinion of the journal and that a correction would be published in the next issue. This did not happen, either in the next issue, or in the following four. Therefore on April 26, 1982 I wrote a letter to Marion Dönhoff, the editor in chief of *Die Zeit*, in which I expressed my surprise about this silence and asked her whether the journal shared Haaf's point of view. I enclosed copies of critical articles about the Unsöld paper, which Haaf had ignored, and mentioned the famous comparison between Einstein and Hitler made by Unsöld in Marburg. A few days earlier Brandt had also written to Dönhoff: In a long and bitter letter of complaint under the heading *Anti-Semitism in Die Zeit* he characterized Unsöld's article as "strongly anti-Semitic" and "anti-constitutional." He mentioned (without citing my name) my leaving under protest the conference hall after Unsöld's 'Hitler = Einstein' sentence and asked for a correction of Haaf's article, for which he considered her responsible.

In her answer of April 19th to Brandt Dönhoff stated that she found his complaints "fully justified" and that she had read Haaf's article only after receiving complaints about it. She also stated that she had asked Haaf to answer the criticism, which she "shared." This was in a certain sense reassuring, because it showed that we were not the only ones who had taken offence. In her answer to me on April 28th she did not express herself regarding Haaf's article, just mentioning that it was up to Haaf to react. In his "reaction" of May 10 Haaf claims that he has done some "research" about the Unsöld affair and got the impression that the reactions to Unsöld's paper were not spontaneous, but "maneuvered from outside" ("why no letters of protest to *Physikalische Blätter* or to Unsöld himself" Haaf asks) and that Brandt and I "try to exercise pressure from behind the scenes." Last but not least, he mentions that he had talked to Unsöld, to Walcher who had chaired the colloquium, and to another participant at the colloquium from the Marburg Physical Chemistry Department, Luck.[c] None of them could confirm that Unsöld had pronounced the famous sentence.

Even more disappointing was the fact that Dönhoff, in a sudden change of mind, apparently adopted Haaf's point of view: In a letter of May 24, 1982 addressed to Brandt she wrote that she initially had trusted his judgment about the anti-Semitic character of Unsöld's article, but that Haaf had convinced her that this was not the case.

I met Walcher some months later in the yard of the Physics Department. He addressed me and mentioned the Unsöld incident. He did not try to deny anymore that sentence and had a possible explanation why he did not remember it: during the colloquium he had left for a few minutes the hall and in hindsight he assumed that Unsöld had pronounced this sentence during his absence. It was my turn not to remember. He asked me why I had been so upset about the whole story: had I, personally, suffered in particular during the

[c]By "coincidence" Luck was one of the publishers of *Physikalische Blätter* and had been, together with Hanle, personally involved in the publication of Unsöld's article. Luck's admiration for Unsöld's philosophy is clearly expressed in a review published subsequently in *Physikalische Blätter* 37 (1981) 329 of Unsöld's book "Evolution of cosmic, biological and mental structures," the very title of his Marburg colloquium. Here are some quotations from this review: "...a unique book...the work and its author deserve everybody's admiration...it is highly recommendable." *Physikalische Blätter* had no problem in publishing this review, after all what had happened.

war? He appeared to be surprised when he heard that I had barely survived the holocaust and that I had lost my mother and many members of my family during that time. He then made a point in emphasizing that he was member of the Social Democrat Party and that he had never sympathized with the Nazi regime. Actually, I have no reasons to doubt that and I believe that his attitude in the Unsöld affair was at least in part dictated by his embarrassment that Unsöld had been his invitee.

In the mean time broader circles got involved in the dispute, among them the President of the Federal Republic, Walter Scheel,[146] who had been personally criticized in Unsöld's article (referring to a speech Scheel gave at Einstein's birth anniversary Unsöld writes "it might be desirable that German politicians, too, should not pity only the emigrants..."), Robert Kempner,[147] the American Prosecutor at the Nürenberg Trials, the Jewish weekly *Allgemeine Jüdische Wochnenzeitung*, which published two very critical reports[148] on this issue, and the *Deutscher Koordinierungsrat der Gesellschaften für Christlich-Jüdische Zusammenarbeit*, the German coordination council of the Christian-Jewish Association. This council represented by Surkau, a senior renowned professor of theology and Stöhr, a minister and president of the association, in an official letter of May 3, 1982 to *Die Zeit*, which the weekly, however, did not publish, after summarizing what had happened at the Unsöld colloquium, asked that the journal reported also other aspects of the Unsöld affair, besides the "censorship" in *Physikalische Blätter*. After expressing their strong disappointment about the missing condemnation of Unsöld's statements by the President of Marburg University and by senior professors of the Physics Department, the authors of the letter point out the existence of a "closed front" of primarily senior professors against Brandt and some junior professors, and that this engagement by the junior professors "will not be helpful for their career."

This sentence alluded presumably to what had just happened: Brandt's and my contracts with the Ministry of Research, through which part of our research projects were funded, had not been renewed. I don't remember what the official motivation in Brandt's case was, but that in my case was so transparent, that an objective observer could hardly have doubts about what was behind.[d]

Theoretical research projects used to be supported by the Ministry of Research only if they were related to experimental programs funded by the Ministry. My previous projects had to do with the nuclear physics experiments of the Gesellschaft für Schwerionenforschung (GSI) in Darmstadt. In 1981 when my projects were up to renewal I wrote a proposal related to heavy ion experiments at CERN. The answer I got, contained, as usual, a part referring to the scientific merits of the proposal and to my previous work. This part was quite positive mentioning that the proposal was worth being supported. When it came, however, to the deeds, the letter regretted that the Ministry could not support the proposal because the experiments at CERN were not funded by it. My surprise was enhanced by the fact that I was certain that this formal motivation did not hold. Indeed, I knew at least one German heavy ion physics group, that of the University of Münster, whose experiments at CERN were funded by the Ministry. I asked the leader of the group to confirm this to me in writing, which he graciously did. I then wrote a strong letter of protest to the Ministry, pointing out that the Ministry apparently did not know what experiments it was funding. I asked to be received by the nuclear science committee, which was handling the grants, to make my point. This happened a few weeks later at the GSI in Darmstadt. What I remember of that meeting is the intervention of one of the members of the committee, who happened to be Brentano and whom I had not met before. He seemed to be very much annoyed by what had happened and asked the chairman some unpleasant questions.

---

[d]In a letter of June 2, 1981 Larry Schulman, a colleague of mine from Indiana University (at present at Clarkson College, New York) informed me that during a recent visit in Germany he "heard my name on a controversial matter. There has been less than overwhelming joy at your having publicized the article by Unsöld....The letter continues: "the reason I'm writing is to tell you I definitely think that you did the right thing....I think you were right to bring the Unsöld article out in the open, if for no other reason than to make the well-meaning people aware that the issue has not disappeared."

Eventually the committee had to reverse its decision and my project was approved.

At the end of 1982 Alfred Grosser, the prominent French sociolog and historian, at that time director of the Institute of Political Sciences of Paris, gave at Marburg University a public lecture on "Germany seen by the French." Grosser has been for many years *the* French expert for Franco–German relations. He used to be consulted by the French government and has been a frequent speaker on French and German TV.

In his lecture attended by an audience of several hundred people Grosser acknowledged the big progress achieved in the democratization of post-war Germany. At the end of his talk, however, he mentioned that what had happened at Marburg University, when a lecturer compared Hitler with Einstein, found little understanding in France. Nobody in the audience, including the University president who was sitting in the front row,[e] commented or contested this statement.

[e]After the talk the president introduced me to the speaker.

19

From Hot Spots to Solitons — On Revient Toujours au Vieil Amour

As always, this time, too, plunging into physics helped me overcome these not so amusing happenings. And according to the above quoted French saying, sooner or later you come back to the old love although in the case to be reported below the love was not so old, after all.

In the discussion of hot spots it was pointed out that the observability of hot spots depends on the relative stability of the excited region against decay. One way of describing this stability was in terms of heat conductivity: The smaller the heat conductivity the longer the life time and the more stable the hot spot is. Heat conduction is from a certain point of view a particular case of hydrodynamical motion, but there exists another hydrodynamical mechanism, which, under certain conditions, could assure stability as well. Such a mechanism is represented by solitons, known for a long time in fluid mechanics.[149] Loosely speaking solitons are stable pulses. They emerge, theoretically, under certain boundary and initial conditions, as time-independent, propagating, solutions of the equations of hydrodynamics.

Given the phenomenological success of the hot spot model, at the beginning of 1980s Gerald Fowler, Sibaji Raha, Norbert Stelte and myself asked whether hydrodynamics wouldn't admit such soliton solutions also for boundary and initial conditions appropriate to the formation of hot spots in nuclear matter. This collaboration came about when, in 1980 Sibaji, who had done his PhD in Austin, Texas on thermodynamics and hydrodynamics of superdense matter, joined the Marburg group, and Norbert Stelte, my first PhD student in Marburg had just finished his thesis on hot spots. (As mentioned before Gerald had been quite a frequent visitor in Marburg where he spent also a sabbatical).

Density Solitons

The history of applications of hydrodynamical methods to nuclear physics had, for various reasons, its ups and downs. After its brilliant start in the 1930s with the Bohr droplet model it took more than twenty years until Glassgold, Heckrotte and Watson again used hydrodynamics for the study of nuclear reactions.[150] This hiatus was due to the big success of the nuclear shell model, which, at a first look, seemed to suggest that the mean free path of nucleons in nuclei is too big to permit the application of hydrodynamics. Glassgold et al. did not accept this argument, challenging the very concept of *collision between two particles* in a strongly interacting nuclear medium. This reminds me of a similar challenge, six years earlier, albeit at high energies, by Landau. In his theory of multiple production Landau argued that in a strongly interacting medium at high energies, the concept of *number of particles* does not make sense but at the end of the reaction, because in the intermediate stage particles are continuously absorbed and reemitted and that, therefore, such a medium must be treated by the methods of continuum mechanics. The 1970s and 1980s brought a resurrection of hydrodynamics both in particle and nuclear physics due to the fact that experimental data from new particle and heavy ion accelerators became available.

Until 1982 the application of hydrodynamics to nuclear reactions had been mostly limited to energies higher than 20 MeV per nucleon, which correspond to supersonic velocities.[151] Glassgold et al., in particular, applied the equations of (non-relativistic) hydrodynamics to the motion of a supersonic hadronic projectile through a large target by expanding the density around its equilibrium value. Since they treated the problem in analogy to the study of a point disturbance in an infinite body of fluid, they kept only the first order terms of the expansion and studied the formation of shock waves, which develop at velocities strongly exceeding the velocity of sound.

We addressed a different problem: We were interested to see what happens *around* the velocity of sound. In particular we studied the motion of a light *nucleus* in nuclear matter, at velocities only slightly exceeding the speed of sound. This put a condition both on the size and on the energy of the projectile:

— its size should be big enough to produce an appreciable change in the density, but sufficiently small (in comparison with that of the target) to be absorbed by the target;

— its energy should be sufficiently big, so that its velocity exceeds the speed of sound, but still small enough to prevent that it "drills a hole" in the target and gets out.

Since we expected a larger density disturbance than in Ref. 150 we repeated the calculation of Glassgold et al., but kept also the second term of the expansion. And, to our surprise, in this way we no longer got the simple, linear, second order wave equation of Ref. 150 but the non-linear, dispersive Korteweg-de Vries (KdV) equation, which is of third order in space-time and which admits, indeed, solitary solutions of density waves. Moreover, the boundary conditions of our problem implied that the solitary density pulse was expected to occur only at velocities slightly exceeding that of sound, which correspond to energies in the range 10–20 MeV per nucleon. This result confirmed our expectation that solitons arise in heavy ion reactions and, given their phenomenological properties, they could be considered as a possible explanation for the formation and relative stability of hot spots in nuclear reactions.

This was not the end of the story but rather it's beginning, because from this moment on the events precipitated and another, this time unexpected, application of solitons emerged. Our first paper on solitons was submitted for publication in *Physics Letters* on April 1982 and appeared in September 1982.[152] Between these two dates, in the June 28th issue of *Physical Review Letters* appeared a paper by Galin et al.[153] in which they reported evidence for a limitation of linear momentum transfer in carbon induced reactions at a bombarding energy of 15 MeV per nucleon. The authors had no explanation for this surprising observation. It did not take us more than a few days to correlate this observation with our soliton effect. We wrote immediately a preprint and in August 1982 we submitted to *Physical Review Letters* our interpretation of the experiment by Galin et al. We pointed that if one accepted the idea of soliton formation in nuclear collisions, the observation of Galin et al. was easily explained: In the reaction the projectile is completely absorbed by the target and induces a solitary pulse, which is moving through the nuclear medium. Since our calculations showed that this soliton couldn't move with a velocity much larger than that of sound, the limitation of longitudinal momentum transfer at 15 MeV per nucleon followed. This experiment constituted thus the first possible evidence for solitary density waves in nuclear reactions. Our paper appeared in the February 7, 1983 issue of *Physical Review Letters*,[154] but the Galin et al. group had noted the preprint already. I was invited to Orsay to present our results there and in a review entitled "Do we actually have strong motives for studying nuclear collisions in the energy range 10 to 100 MeV/A" Marc Lefort, the leader of the group and builder of the GANIL accelerator designed to cover that energy range, so critical for the appearance of solitons, quoted this effect as one of the reasons for the forthcoming investigations in this energy range.[155] Encouraged by these follow-ups, we wrote two more papers in which we investigated the stability of solitons in three dimensions[156] and the influence of the nuclear surface on their motion.[157]

From Jupiter to Atomic Nuclei — Solitons in Rotating Nuclei

After our incursion into hydrodynamics of density waves, which lead us to the prediction of *density* solitons in heavy ion reactions, we asked ourselves[158] whether this analogy between macro and microscopic physics, inspired by the phenomenon of hot spots, cannot be used for rotating systems as well. The answer again turned out to be affirmative.

In macroscopic, classical physics hydrodynamics predicts certain effects, which arise only in the presence of rotation. This time the starting point of our analogy was the macrocosm itself and in particular the Great Red Spot (GRS) of Jupiter. Only a few years earlier Maxworthy and Redekopp[159] had suggested that this Red Spot effect was of hydrodynamical origin and due to specific solitary waves, known in the literature under the name of *Rossby waves* and which develop in the Jovian atmosphere as a consequence of the rotation of the planet. In this interpretation the GRS is a localized excitation on the surface of the planet and the solitary character of the Rossby waves assures its stability. The possible analogy between the GRS and the hot spot effect in particle and nuclear physics was tantalizing. This time, however, we started from another, more direct link, between microphysics and hydrodynamics: The equations of quantum mechanics, in particular the Schrödinger equation, can be written in a form identical with that of the equations of hydrodynamics. This fact had been known since 1926[160] and had even been used in the 1970s in nuclear physics, albeit in a different context. In this formalism we are dealing from the start with *quantum hydrodynamics* and *probability fluids* rather than classical hydrodynamics and nuclear or hadronic fluids. Here from follows that in the applications to subatomic physics no assumption of local equilibrium is necessary. While for multiparticle production processes in particle and nuclear reactions this method is, because of practical reasons, not very useful,[161] for the particular problem of rotating hot spots it is.

The key observation of Ref. 158 was that, as a consequence of two-particle correlations, the velocity field corresponding to the many body system represented by the nucleus was rotational. Mathematically this implied that for rotating nuclei one had to respect the conservation of potential vorticity[162] relative to global rotation and this lead to the appearance of *quantum* Rossby waves. By treating these waves similarly to the procedure of Ref. 160, we eventually found a localized vortex-like velocity field, which was concentrated on the surface of the system and which was different from that associated with rigid rotation. Applied to deep inelastic heavy ion reactions this finding implied that the hot spots observed in certain experiments might have, as in the case of Jupiter, a hydrodynamical origin: in the final state the reaction partners, after separation, rotate and on the surface appears a local velocity field, the magnitude of which is different from that of the rigid body. Our paper[158] concludes with the sentence: "If these vortices are indeed the origin of 'hot spots' then they would not really be 'hot,' although containing localized energy."

This analogy with planetary physics and exchange of ideas between truly macroscopic and microscopic physics lead also to an exchange of the respective proponents, when Tony Maxworthy visited me in Marburg and I visited him at the University of Southern California in Los Angeles.

20

Caught Up Again by the Past

The Einstein Scandal: Follow-ups 1983–1989

Kiel: Albrecht Unsöld is named honorary senator of the University of Kiel.[163]

Frankfurt: Brandt publishes in the monthly journal *Tribüne*[164] an article about the Unsöld affair under the title "Atombombe — Schuld der Juden?" (The Atomic Bomb — Are the Jews to blame?)

Wiesbaden: On December 10 1984, in a session of the Parliament of the Land of Hessen the deputy K. Schilling mentions the Unsöld colloquium as an anti-Semitic incident quoting, amongst other things, the comparison between Haber and Einstein, and Hitler.

Marburg 1987–1989: The Land of Hessen, being the host of the Heavy ion accelerator in Darmstadt, created a special fund to support research on this topic at the universities of Hessen. Somehow, however, this initiative was not publicized, so that I learned about it only through Brandt, after seven professors from the Physics and Physical Chemistry Departments, including him had written a report about the work on nuclear and heavy ion physics at the University of Marburg. This report was the necessary condition to get support. Strangely enough, I was the only professor working in this field, which had not been invited to contribute and whose work was not mentioned in this report, although my publications on this subject were a matter of public record. Actually, some of the contributors were beginners in this field and I had even introduced one of them into that subject. Brandt immediately related this incident to the Unsöld affair and I, reluctantly, had to agree with him on his point. He, however, went even further and made it public, considering it a manifestation of anti-Semitism. When interviewed by the local newspaper *Oberhessiche Presse*, I myself characterized it rather as a mixture of stupidity and attempt to mislead. It must be said that the (new) president of the university, urged by Brandt and others to put the record straight, not only refused to do that, but in a letter addressed to me reproached me for "not having opposed the accusation of anti-Semitism," which I actually had done in the interview quoted above. Only after the story found again a strong international resonance — many people saw in it a new attempt of revenge for making the Unsöld story public — the physics department in a face saving operation officially criticized the initiators of the incident. This time the president went out of his way to "appease" me; he sent me a handwritten letter congratulating me for my election as a member of the New York Academy of Sciences. Actually this had happened 11 years earlier!

At the end of 1988 I had a heart attack and had to undergo heart surgery. After returning from hospital I was visited at home by the university president who, like many others, rightly saw a connection between this health event and the events preceding it.

One might be amazed by the astonishing ignorance, confusion and lack of sensitivity of certain German scientists and journalists and by the realization that hadn't be the reaction from abroad Unsöld might have continued to find followers. (Indeed the historian of science will notice that among those who condemned Unsöld's statements were, mainly, either foreigners, or Germans who during or after the war had lived abroad[165]). However the reader is advised to keep his amazement reserved for the next story.

The Auschwitz "Experiments"

Another series of events, which shows the difficulties Germany met in coping with its past is represented by the history of the Max Planck Gesellschaft. I will briefly mention this history here, because these events happened twenty years *after* the Einstein-Unsöld scandal and because the dimensions of this story go much beyond those described above.

However unbelievable it might appear today, Mengele and collaborators carried out their "experiments" on human beings of "inferior race" under the auspices of the Kaiser Wilhelm Gesellschaft (KWG). This research organization comprized all major scientific research institutes of Germany, including that at which Einstein and Max Plank worked. In 1948 the name of this organization was changed into Max Planck Gesellschaft (MPG). Initially the connection between the KWG (represented among others by Butenandt, the director of the Berlin Biochemistry Institute and Nobel Prize laureate) and these "medicine experiments" was frantically denied. Later on, in view of the irrefutable facts proving this connection, another strategy of interpreting this link was adopted: the "experiments" got a scientific justification! Moreover, the authors of this reinterpretation were not certain politicians or historians, but the Max Planck Gesellschaft itself.

In an official report (of 1000 pages) on the history of the KWG "*Forschung im Spannungsfeld von Politik und Gesellschaft*" (Research between the poles of society and politics) issued in 1990 the following statement about the Nazi racial laws can be found[166]: "The new laws were beneficial not only to the 'entire nation' (*Volksganze*) but also to science. The institutes of anthropology, human genetics, eugenics and psychiatry of the KWG got new tasks in science, teaching, research and refereeing, beyond that of fundamental research." According to this report even mass murder became a useful research topic: "The mass killing of mentally ill opened new research possibilities also for the brain anatomy section of the KWG." The author of the article,[166] as well as others before him, asked the President of the Max Planck Gesellschaft to apologize in the name of the KWG for these crimes. The president refused to do that,[167] arguing that the work of the commission, which investigated these crimes, was not yet finished ("At first investigation, then action"). Moreover, for somebody like him, who had not been personally involved in these crimes, such an apology would be an act of "presumptuousness." What is even more disturbing, he did not find it necessary to distance himself from the unbelievable statements of the report quoted above; he did not mention them with one single word. Was it necessary to wait more than fifty years for the definitive verdict of a "commission" to understand what had happened? It is interesting to mention that some of the heirs of the perpetrators had a different attitude. Michael Wörle, the grandson of the Director of the KWG Institute mentioned in Klee's article has dedicated his life to the investigation of the Auschwitz "experiments" and in a letter to *Die Zeit* (March 2, 2000) criticizes Markl's refusal to ask the last survivors of the Auschwitz experiments for forgiveness. Wörle states: "Scientific clarification is one thing, apology another."

The presumably last act of this tragedy reminded me of the Unsöld affair. In 2001, as a result of increased pressure from abroad (and *before* the final "conclusions" of the commission were published) the same president of the MPG finds, after all, the "courage" to act against his own convictions and to apologize, in the name of the KWG and its successor, the MPG, for the crimes committed in the name of science.[168]

This sad story made me recall what Einstein wrote on January 28, 1949 to Otto Hahn, when he explained to him why he could not accept to become an external member of the MPG.[169] Referring to what had happened during the Nazi time Einstein wrote: "The attitude of the German intellectuals... was not better than that of the mob. One does not see even remorse and a honest will to repair the little that still can be repaired after the terrible murders."

I wonder what Einstein would have thought had he known the *history of the history* of the organization that he had been a member of until 1933.

Section VIII

From Optics to Particle Physics; Traveling to the East

21

Quantum Optical Analogies and Methods in Strong Interactions High-Energy Physics

The link between optics and particle physics has two aspects — the phenomenological one and the formal one. The first manifests itself in certain phenomena, which are common to both fields (interference, diffraction, transparency, etc.) The formal link is based on the fact that both in optics and particle physics one applies quantum field theoretical methods.

High-energy physics is particularly close to optics when identical particles are considered, since in optics, obviously, only photons are involved. What distinguishes optical phenomena from those in high-energy physics are conservation laws and final state interactions, which are present in high-energy physics, but not in optics. (Final state interactions are interactions between particles produced in the final state and which usually are not considered in the production mechanism. Examples are hadronic resonances or electromagnetic interactions between charged particles).

At high energies and high multiplicities the first are unimportant. Under certain circumstances final state interactions can also be neglected and then we may take over the quantum optical formalism and use it for the interpretation of multipion production at high energies, provided we consider identical pions. A review of the applications of quantum optical methods to multiparticle production can be found in Ref. 170.

I will try to sketch below the personal circumstances that lead to my involvement in the application of quantum optical methods to particle physics. I mentioned already that my preoccupation with optics goes back to the fifties when I came across the isomeric shift on spectral lines. That raised my interest in intense sources of light so that I followed the invention of masers and lasers even before the development of the formal theory of quantum optics. After a pause of about a decade, my interest in this subject was again stimulated in the seventies by certain developments to be enumerated below, all of which have to do with the notion of coherent states.

Coherent states are the alpha and omega of quantum optics (QO). This is so because coherent states are the eigenstates of the annihilation operator, which in plain language means that by taking out one particle from that state, one recovers the same state.[171] This implies that the number of particles in a coherent state is undefined. But this is just the situation we face in quantum optics, where we deal both with large and small number of photons and where in general this number does not play a role. Instead, it is the canonical conjugate of the number of particles, the phase, which is important.

The developments, which determined my interest in coherent states, in particular are the following.

— Gauge theories: After the successful unification of electromagnetic and weak interactions in the 1960's, the early 1970's witnessed the establishment of the theory of strong interactions in the form of quantum chromodynamics and what is called today the standard model became part of the accepted lore of theoretical particle physics. The electroweak theory and QCD are gauge theories[172] in which the vacuum, the state with no particles, plays a special role: its symmetries are different from that of the underlying theory; the symmetry of the theory is spontaneously broken. But the vacuum is a coherent state.

— Classical solutions of field theories: Certain field theories, which are of interest for strong interactions admit classical solutions, e.g. solitons (cf. also chapter 19). Classical fields are also coherent states.

— My work on superfluidity of hadronic matter: this work raised my interest in Bose-Einstein condensates, i.e. states made of bosons where all particles "condense" into one single state. These condensates are again coherent states. This last fact as well as my previous preoccupations with the "mysteries" of quantum physics (cf. chapter 31) related to the principle of identical particles, which is reflected in particle correlations, had made me follow also the developments in the field of Bose-Einstein correlations. In this context I should also mention an analogy between superfludity and the notion of coherence. While the concept of coherence and in particular that of coherent state is of fundamental importance in quantum physics and will be discussed in greater detail later on here it is enough to use the intuitive meaning of coherence, to prove the analogy. Friction arises as a consequence of the difference in momentum loss of individual atoms. In a coherent state there is no such difference; all atoms have the same momentum, they are all condensed in the same state. But no friction means superfluidity. Another facet of this relationship is represented by the energy gap: since a superfluid is a coherent system (the exact statement is that the condensate is a coherent state) it is "rigid," hence it cannot be easily rotated. Here from follows the existence of the characteristic energy gap.

To all these factors another important personal factor has to be added: in the early seventies I met Gerald Fowler, at that time the professor of theoretical physics at the University of Exeter. Gerald got interested in my work on superfluidity of hadronic matter and during my stay in London in 1972–1973 he invited me to Exeter to give a talk on this subject. None of us would have thought that this would be the starting point of a collaboration and a friendship, both of, which would last almost twenty-five years until his death in January 2000. After I joined (in 1974) the University of Marburg, Gerald was a frequent visitor there and we published together an appreciable number of papers on Bose-Einstein correlations and other applications of quantum optics in particle and nuclear physics.

Gerald, nine years my senior, had done his PhD with Rosenfeld in Manchester and had participated in the historical developments of particle physics in Manchester and then in Bristol, before accepting the chair in Exeter.

As I understood from Gerald, and this was corroborated by what I heard from Jacques Prentki and Roberto Salmeron, who had also been in Manchester at that time, the period of the late forties and early fifties in Manchester must have been quite fascinating, since at that time Britain was the center of particle physics in Western Europe. It was the only Western country which had not been occupied during World War II and which had participated, together with the USA, in such important joint scientific ventures like the atomic bomb, radar and cryptography. That is why soon after the war it became the Mecca of physicists from all over Europe. I remember an amusing story Gerald told me once about his relationship with Rosenfeld. As any scientific institution the physics department in Manchester was confronted, from time to time, with "revolutionary inventions" and, as part of the social obligations of a university institution, it had to judge the scientific merits of these proposals. One day Rosenfeld asked Gerald to referee about such a proposal (if I remember correctly it was a precursor of the later fork bending "experiment" of Uri Geller). Gerald, as a young PhD student felt honored by the confidence his professor had in him and took this task very seriously. He spent several days analyzing all the details of the proposal and then wrote a long report in which he concluded that the "proofs" were not convincing. To his amazement Rosenfeld did not even look at the report and just said: "What else did you expect?"

In Exeter Gerald was supposed to do research in those fields of physics in which his experimenter colleagues worked, and these topics did not include particle physics. Physics and science in general went in Britain already through the "fasting" period, which had begun a few years earlier in the US. In Exeter this meant, among other things, that the library ceased to order particle physics periodicals like *Nuclear Physics*. Gerald adapted himself quite quickly to the new situation and started to study the quantum optical literature, a topic that had survived. Still, being a person of broad scientific interests he continued to follow what was going on in particle and nuclear physics by reading the *Physical Review*.

Given the circumstances enumerated above it appears now, in hindsight, quite natural that after some discussions and correspondence we two should ask ourselves whether the coherent states, which apparently were present in strong interactions would not manifest themselves in Bose-Einstein correlations of hadrons, as well, reflecting the coherence of the emitting source. In lasers, which are due to coherent sources, this effect was well established, but the analogy between coherence of light sources and that of hadronic sources was limited by the presence of final state interactions and conservation laws, which affect correlations between hadrons and which are absent in the case of photons.

At this point I should also mention my encounters with one of the founders of quantum optics, Roy Glauber, whom I think I had met for the first time in 1969 at CERN and whom I visited regularly later on at Harvard, whenever I passed through Boston, which happened quite frequently. Strangely enough, in my meetings with Glauber we discussed various subjects of common interest, mostly collision theory in nuclear reactions, but, as far as I remember, the topic of Bose-Einstein correlations was never mentioned in these discussions. As a matter of fact, Giuseppe Cocconi with whom I discussed this subject in the late 1970's told me that he had tried in vain for years to get Glauber again interested in this subject, and in particular in the applications of this effect in particle physics. And this despite the fact that Hanbury-Brown and Twiss intensity interferometry had been the starting point of Glauber's pioneering work on quantum optics. However when we met again at the end of the 1990's in Trento at a workshop dedicated entirely to this topic I was pleased to find out that Glauber not only knew about the work of Gerald and myself, but also about some later work of the Marburg group. In the following I will discuss those common topics of quantum optics and particle physics in which I was involved. These include: Bose-Einstein correlations, multiplicity distributions, self-induced transparency, superradiance, superfluorescence and squeezed states, and cover a period of three decades.

22

Bose-Einstein Correlations

The physical effect of interference takes place when waves are superposed and is thus common to all chapters of physics where waves appear: hydrodynamics (fluid waves), acoustics (sound waves), optics (light waves), and quantum mechanics (probability waves). The concept of wave originated in fluid mechanics where from it was taken over by analogy to other fields. The interference between two (or more) waves establishes a *correlation* between these waves. In particle physics, in particular, where to each particle there is associated a wave, we encounter thus interference and correlations between two (or more) particles, described mathematically by second or higher order correlation functions.[173] The different behavior of bosons and fermions manifests itself in a very direct way also in the phenomenon of correlations of identical particles. We thus distinguish Bose-Einstein and Fermi-Dirac correlations. While in Fermi-Dirac correlations the particles are antibunched, in Bose-Einstein correlations they are bunched.

Intuitively the anti-bunching effect can be explained as a consequence of the Pauli repulsion. Mathematically bunching and anti-bunching are described by the value of the correlation function at the point where the momenta of the two particles are equal; this value is called the intercept.

Besides bunching, Bose-Einstein correlations have also another property, which is not found in Fermi-Dirac correlations, namely coherence. This property leads under appropriate conditions to Bose condensation, an effect, which is at the origin of such important phenomena as superconductivity, superfluidity, masers and lasers. It is worthwhile reminding that the phenomenon of condensates was discovered before its "parents," i.e. before Bose-Einstein correlations.[174]

The notions of interference and coherence originate in classical optics. Here two beams of light are said to interfere coherently when the phase difference between their waves is constant; if this phase difference is random or changing the beams are incoherent. The concepts of interference and coherence have evolved through analogies. Thus up to the middle of the 20th century, coherent interference was limited to the coherent superposition of wave *amplitudes*. That is what is called today amplitude or first order interference. In analogy to that, Hanbury-Brown and Twiss (HBT) have introduced in the 1950's the concept of *intensity* or *second order* interference, which generalizes the interference between amplitudes to that between squares of amplitudes, i.e. between intensities.[175]

Amplitude interferometry is used for the determination of lengths, surface irregularities and indexes of refraction. With intensity interferometry, besides measurements of distances (and lifetimes, in the case of particle physics) another application has become possible, the determination of quantum statistical — i.e. higher order — coherence of sources.

The HBT intensity interferometer, which is the first realization of higher order interference, has brought important progress in the technique of determination of sizes of stars. However, intensity interferometry has brought also much more, it lead to the development of a new chapter in physics, that of *quantum optics*, the discipline, which, among other things, explains, within quantum mechanics, how such important devices like masers and lasers work. Thus the evolution from amplitude to intensity interferometry also corresponds to the transition from classical to quantum physics. This evolution proceeded through the generalization, in 1963, by Glauber, of the concept of coherence, by introducing the concept of higher order coherence. Maximum first order coherence is achieved when the average of the product of wave amplitudes equals the product of the averages of the amplitudes, i.e. the amplitudes factorize. In analogy to that second (higher) order coherence is defined by the property that the squares (higher powers) of the amplitudes factorize.

However, between the two important events in optics mentioned above — the invention of intensity interferometry due to HBT in the 1950's and the formulation of quantum optics by Glauber in 1963 — another event, which appeared initially unrelated, took place in particle physics. In 1960 Gerson Goldhaber, Sulamith Goldhaber, Wonyong Lee, and Abraham Pais (GGLP) discovered[176] that identically charged pions produced in antiproton-proton annihilation processes were bunched, while pions of opposite charges were not. They correctly interpreted this effect as due to Bose-Einstein statistics. At that time GGLP were apparently not aware of the relationship between their effect and the HBT experiment (this relationship was established only much later by Grishin, Kopylov and Podgoretskii.[177] In fact, the two experiments deal with the same physical effect, because the intensity interference measured by HBT is also due to Bose-Einstein correlations, correlations of photons. Although Hanbury-Brown and Twiss as well as others, including Glauber, were aware of the connection between intensity correlations and Bose-Einstein statistics, they, apparently, did not relate the HBT and the GGLP effects either. A possible explanation for this fact is that intensity correlations can be formulated also in classical physics terms, as long as the nature of the particles involved is not manifest. This applies in particular for HBT interferometry, where one deals, effectively, with intensities of light and where the number of particles (photons), out of which the intensities of light are made, is large and not counted. In the GGLP effect, however, the quantum nature of the effect is obvious, since the pions, which are bosons, are counted. As a consequence of this initial separation between quantum optics and particle physics, it took almost 15 years until the theory of Bose-Einstein correlations in particle physics benefited from the important developments that had taken place in quantum optics. I refer in particular to the concept of coherence.

It was mentioned already that in quantum optics coherence plays a central role. Given the quantum nature of the phenomenon of Bose-Einstein correlations, it appears quite natural to expect that coherence should play an important part in that effect, too. However, what appears now, in hindsight, natural was not so in the past. While the first experimental observation of intensity interferometry in particle physics by GGLP goes back to 1959–1960, our observation that coherence should manifest itself also in higher order *particle* correlations and that there exists possible evidence for this effect in the data was presented by myself as late as 1977 at the Kaysersberg Multiparticle Dynamics meeting and published in the same year also in Ref. 178. Actually, the "public opinion" was not yet prepared for this realization. The Kaysersberg meeting was organized by the Nuclear Physics Institute in Strassbourg and Saul Barshay had to do some tough convincing work to persuade the organizing committee of the meeting to invite me to that meeting and to present our arguments. It happened that during that period Saul Barshay, whom I had met several years before at the Bohr Institute in Copenhagen, and who belonged in 1977 to the Strassbourg institute had just published in collaboration with Arnold a paper emphasizing the absence of experimental proofs for the existence of coherence. I remember that meeting among other things also because I met there for the second time Feynman, who again impressed me by his friendliness and accessibility. The first time I had met him at a meeting in the States and had a short discussion with him about superfluidity of hadronic matter.

It must be mentioned that to prove experimentally coherence in high-energy physics Bose-Einstein correlations has been quite a difficult task, because these correlations are rather insensitive to even large admixtures of coherence and also because of other competing processes, which could simulate this effect.[179] However the importance of this phenomenon was recognized almost immediately, once a new aspect of coherence was put in evidence[180]: the relation between coherence and the classical fields present in the standard model. To the publicity for the coherence effect contributed also an important paper by Gyulassy, Kaufmann and Wilson published around that time[181] in which the realization about the possible role of coherence in BEC was characterized as "the most exciting recent theoretical development." At present almost all publications on Bose-Einstein correlations, both of experimental or theoretical nature, discuss this effect.

Still, until the end of the 1980's, even those experimentalists working on BEC who had been aware of the problem of coherence, used an inadequate formalism in the interpretation of their data, although the correct quantum optics formalism was available in the literature. I realized that during my hospital stay in 1989, recovering from heart surgery, when I had more time to think about physics and life and had "escaped" for a while from the pressure of writing proposals to be able to pay my collaborators. I then wrote a short paper[182] pointing out this case of collective amnesia.

Attempts of Falsification

Though this be madness, yet there is method in't. (Shakespeare, Hamlet)

According to Popper, a theory deserves the epithet "scientific" if it resists attempts of falsification. From this point of view the quantum optics approach to BEC has passed with brio this critical test. Indeed, one of the most convincing experimental confirmations of quantum coherence came from the measurement of higher order correlations in antiproton-proton reactions at the CERN SPS collider by the UA1-Minium Bias collaboration.[183] These predictions include among other things the relation between first, second and higher order correlations. This experiment has a particular significance because it tests in quite an unusual way the predictions of quantum optics as applied to BEC: it represents an involuntary attempt of falsification of the theory.

While quantum statistics[a] predicts the relationship between various orders of correlations it does not predict the analytical form of the correlator, i.e. the first order correlation function, which describes the transition amplitude between two states of one particle and which has to be determined experimentally. The correlator contains as a parameter the radius R of the source, which is, besides the amount of coherence λ, the main physical quantity we are after in interferometry. With a given form of the correlator the relationship between first and higher order correlation functions is uniquely prescribed by quantum statistics and in this relationship also influences the value of the coherence parameter λ. By experimentally determining the form of the correlator and then by measuring the correlation functions one can determine R and λ. An important test of the theory is the fact that the same parameters R and λ appear in all orders of correlations.

At the time the authors of Ref. 183 made their measurements there existed in the literature predictions (based on quantum optics) of higher order BEC for a Gaussian correlator only. However the data of Ref. 183 could not be fitted by a Gaussian correlator, but rather by an exponential one. To proceed further the experimentalists used the available higher order correlation formulae, derived for a Gaussian correlator, for the exponential correlator as well. The results of this procedure, which were immediately published in preprint form and then in *Physics Letters*[183] were somehow surprising: the data for the various orders n of correlation functions could not be fitted by the same two parameters R and λ, as implied by quantum statistics. However, in this apparent disagreement there was a systematic pattern, as if illustrating the quotation from Shakespeare's drama: while λ was

[a]Here and in the following we will sometimes use instead of quantum optics the notion of quantum statistics to emphasize that it applies not only to photons but to identical particles in general.

independent of the order n, R depended on n in a well-defined way. This dependence was called in Ref. 183 "phenomenological."

I learned about these results from Leo Razumov, who had read the preprint and who came immediately to my office, being quite bewildered about what had happened. Leo had realized what the authors of Ref. 183 had done and had immediately checked that the apparent dependence of R on n was exactly cancelled if the formulae used by the experimentalists were corrected in the way prescribed by quantum optics. The Gaussian correlator $\exp(-R^2q^2)$ being the square of the exponential one $\exp(-Rq)$, the change of the relationship between various orders, when replacing one correlator by the other, implies an effective change of R, for each order, explaining the apparent dependence of R on n reported in Ref. 183. By empirically modifying the formulae of higher-order correlations Ref. 183 had explicitly violated quantum statistics and the "phenomenological" relationship derived between R and n[184] just compensated for this violation. Michael Plümer, Leo Razumov and I realized that this accident was much more than a simple error: it was highly significant from a heuristic point of view, because the fact that quantum statistics was effectively "reinvented" by an unbiased experimental team was quite a convincing proof of its validity in particle physics and at the same time evidence for coherence. As far as I remember I wrote immediately to Brigitte Buschbeck in charge with the UA1 Minimum Bias collaboration and told her how happy we were about their results and how important we thought they were.[b] We expressed this also in a paper entitled "Evidence for quantum statistical coherence from experimental data on higher order Bose-Einstein correlations"[185] in which we put the record straight. Even now, so many years after that event, I continue to wonder what other examples of this kind exist in the history of physics and of science in general.

[b]Initially Brigitte whose high competence and qualifications I have always appreciated, felt embarrassed by what she considered a mistake. But I hope that I succeeded in convincing her later, in our frequent meetings, that this "mistake" was actually a great service to science.

Starting with the 1980's BEC has become a "fashionable" subject in high-energy physics and at present meetings entirely dedicated to this subject take place.[186] Another reason for this interest in BEC is that it is, up to now, the only method for the determination of sizes and lifetimes of sources of elementary particles. This is of particular interest for the ongoing search of quark matter in the laboratory. Indeed, to reach this phase of matter a critical energy density (of the order of a few GeV/fm^3) is necessary. To measure this energy density we have to determine the volume of the fireball in which this matter is supposed to have been generated and this means the determination of the size of the source. Here the method of intensity interferometry enters the play. Furthermore, when we talk about a *phase* of matter we have in mind a quasi-stable state, which lives longer than the duration of the collision that gave rise to this state. This means that we have to measure the lifetime of the new system. This lifetime can again be obtained by BEC only.

Bose-Einstein correlations applied to strong interactions can however provide much more: they can be used for the determination of such important dynamical properties like correlation lengths and correlation times.[187] Moreover, besides these practical applications of BEC in interferometry, this new space-time approach[187] has led to quite an unexpected heuristic application, related to the principle of identical particles, the fundamental starting point of BEC.

Some Pions Are More Equal Than Others and Some Unequal Pions Are More Unequal Than Others

To explain this strange result a small detour into quantum mechanics is necessary. As long as the number of particles of a quantum system is fixed, the system can be described by a wave function, which contains all the information about the state of that system. This is the first quantization approach and historically Bose-Einstein and Fermi-Dirac correlations were derived through this wave function formalism. In high-energy physics, however, we are faced with processes where particles are produced or absorbed and this demands a more general field theoretical approach called second quantization. This is the approach on which quantum optics is based and it is only through this more general approach that quantum statistical coherence, lasers, condensates and some quite new phenomena to be mentioned below could be interpreted or discovered.

The wave function of two identical particles is symmetric or antisymmetric with respect to the permutation of the two particles, depending whether we consider identical bosons or identical fermions. For non-identical particles there is no permutation symmetry and at a first look there should be no bunching or anti-bunching. Let us consider however a pair of particles made of a positive and a negative pion. These particles are of course non-identical and according to the wave function formalism there should be no correlation between these particles. Actually, this is the way GGLP discovered their effect. However, this is true only in a first approximation. If one considers the possibility that a positive and a negative pion are virtually related in the sense that they can annihilate and transform into a pair of two neutral pions (or two photons), i.e. a pair of identical particles, we are faced with a more complex situation, which has to be handled within the second quantization approach. This was done in Ref. 188 where it was found that there exists a new kind of Bose-Einstein correlations, namely between positive and negative pions, albeit much weaker than that between two positive or two negative pions. On the other hand there is no such correlation between a charged and a neutral pion. Loosely speaking a positive and a negative pion are less unequal than a positive and a neutral pion. Along the same lines it was found that the BEC between two neutral pions are somewhat stronger than those between two identically charged ones: in other words two neutral pions are "more identical" than two negative (positive) pions.

These results were proven by using the classical current approximation. This method is inspired by an analogy with macroscopic classical physics where the currents are usually external and unchanged by the interaction with the fields. Mathematically they are c numbers and the formalism is then identical with the coherent state formalism and the corresponding equation can be solved exactly. The quantum corrections to this formalism were studied in Ref. 189.

The predicted effects appeared so surprising — the title of the first paper on this subject published in *Physical Review Letters* was actually "Surprises from Bose-Einstein Correlations" — that at the moment when they were obtained some people could not believe them. To quote from one of the first reactions to these results[190]: "It seems unbelievable that Bose-Einstein correlations between neutral pions differ from those between (identical) charged pions and it seems unbelievable that there exist Bose-Einstein correlations between positive and negative pions, which are not identical. Such correlations do not exist between π^+ and π^0."

The reason for this reaction of the scientific community lies in the fact that the naive wave function approach to BEC was still deeply rooted and there is no obvious way to obtain these surprising effects within that approach. At present there exist at least three other derivations of these effects, among them one due to Bowler,[189] and they constitute a definite challenge for experimentalists. The fundamental importance of these predicted effects justifies the efforts necessary for their detection. (The effects are quite small and necessitate high statistics to be seen in experiment).

The surprising BEC effects described above illustrate the superiority of the field theoretical second quantization approach as compared with the wave function formalism. But they also illustrate the limitations of the analogy between optical and particle physics interferometry. Indeed, optical intereferometry deals with photons only and from the above considerations follows that Bose-Einstein correlations between two photons are different from those between two identically charged pions, an issue which had led to misunderstandings in the theoretical literature and which was elucidated in Refs. 191 and 192.

The multitude of problems and aspects of Bose-Einstein correlations, the ever increasing role this phenomenon plays in high-energy and nuclear physics and the large number of investigations dedicated to it (only the Marburg group published more than two dozens of papers on this topic and that made the late Peter Carruthers call me "Mr. Bose-Einstein") convinced me about the need for reviewing the subject. I did this in the paper[191] and in the book,[192] which is the first and so far the only textbook on this subject. Some of the most important papers on this subject were edited in Ref. 193. Most of this work was done in Paris after my official retirement from the University of Marburg.

23

Order and Chaos

Nothing exists except coherence and chaos, everything else is opinion

The studies on Bose-Einstein correlations of the Marburg group were in some sense a by-product of our investigations dedicated to coherence. Another result of these investigations is that related to order and chaos, a topic, which in the last decades became quite fashionable in other chapters of physics, as well.

Coherent states have the remarkable property that their entropy vanishes, which means that they describe, from the statistical point of view *order*. Maximum entropy corresponds to *chaos*, which represents the opposite of order and in analogy to this statistical interpretation the states, which describe in quantum statistics chaos, are called chaotic states. In general one has mixtures of coherent and chaotic states and Bose-Einstein correlations reflect these mixtures or superpositions. While condensates and lasers are described by coherent states, systems in thermal equilibrium are described by chaotic states.[194] Therefore BEC can serve as a tool not only for the measurement of sizes and lifetimes of sources, but also for the determination of the state (of matter) of the system. This is another reason why BEC play an important role in the search for quark matter, since, by definition, a *state of matter* is a state of themodynamical equilibrium.

The phenomenon of chaos is characteristic for non-linear dynamical systems. It manifests itself through a high sensitivity to initial conditions: a small change in the initial conditions can produce a huge change in the evolution of the system. This phenomenon is not limited to physics; it is seen among other things also in economy and population growth. This illustrates the extent up to which certain analogies are applicable. In physics, besides turbulence and meteorology, chaos is present also in quantum optics and through this portal it made its entry into particle physics, too.

From Correlation to Multiplicity Distributions

Besides Bose-Einstein correlations, chaos and its opposite, order, i.e. coherence, manifest by themselves also in another physical quantity, that of multiplicity distributions. Since the *number* of particles produced in a given reaction is not fixed, varying from event to event, it is useful to consider the *probability* to observe in a given event a certain number of particles. The associated *multiplicity distribution* (MD) is one of the most characteristic physical quantities in particle physics, as it allows distinguishing between order and chaos: The multiplicity distribution, which corresponds to coherent states has the mathematical form of a Poisson distribution. Its "opposite," i.e. the distribution, which corresponds to chaotic states, is the negative binomial.

The multiplicity distribution, or more precisely, the mean number, which is determined by that distribution, has also another very important property: it characterizes the type of the interaction. Indeed, for electroweak interactions, because of the smallness of the coupling constant, the mean number of particles is much smaller than unity, while for strong interactions the opposite is true; at present in high-energy particle reactions — and in particular in high-energy heavy ion reactions — events with hundreds of secondary particles are frequently observed.

Although the relation between MDs and correlation functions is a straightforward consequence of multiparticle dynamics and was explicitly derived in Ref. 180, the connection between MDs and correlations has often been overlooked. Sometimes models for one of these two observables were proposed without any reference to (and even in contradiction with) what was known about the other. This is due in part to the fact that measurements of correlations are, for reasons of statistics and other technical considerations, frequently performed in different (narrower) regions of phase space than the measurements of MDs. However, just because of the different experimental methods used in the investigation of these two observables, the relationship between multiplicity distributions and correlations is very important, since the systematic errors involved in each method are also different and a comparison of the results can contribute to a reduction or even elimination of these errors. Therefore, although theoretically MDs and correlations are equivalent, from a phenomenological and practical point of view they are rather complimentary and have to be interpreted together.

The application of quantum statistical methods to multiplicity distributions proceeded in two stages. In the 1970s and 1980s the experience gained in quantum optics was used and some of the results were presented above. In the 1990s, the developments in the theory of Bose-Einstein correlations and in particular the space-time approach to BEC[187] have made possible an interpretation of multiplicity distributions in which the dynamical and space-time properties of the source as seen in BEC are taken into account.

The Time-Rapidity Analogy

In strong interaction phenomenology there exists an observable, which plays a particularly important role, the rapidity. It characterizes the speed of the particle (here from its name) similarly to its cousin, the pseudo-rapidity, introduced half a century ago by Landau, and which characterizes the angle. The historical importance of rapidity lies in the fact that in the central rapidity region one observes a plateau the width of which broadens with energy. Although this flat distribution is most probably an approximation of a Gaussian, a form predicted by the Landau hydrodynamical model,[195] it led us[180] to a useful analogy based on quantum optical considerations, the analogy time-rapidity.

In optics processes are usually stationary in time, i.e. the phenomena do not depend on the particular moment at which they occur. This is true in particular for the optical intensity, which is related to the multiplicity-distribution. A similar phenomenon has been observed in strong interactions particle physics, when replacing time by rapidity. In high-energy reactions the multiplicity distribution in the plateau region does not depend on the particular value of rapidity considered.

In the last two decades of the 20th century the dependence of the multiplicity distributions on rapidity has been in the center of multiparticle production studies. This happened among other things because the large multiplicities produced in high-energy reactions made possible the study of particle production phenomena in *finite* rapidity bins. Actually the study of MDs in finite rapidity bins was initiated in Ref. 196 where we predicted a new effect based on another analogy with quantum optics — self-induced transparency — to be discussed later on. Subsequently the study of MDs in finite rapidity bins became very fashionable, albeit for antiproton-proton and e^+–e^- reactions; only our group published quite a few papers on this topic.

One of the main experimental observations made in multiparticle production was the broadening of the MDs with the decrease of the width of the rapidity bin. While initially this observation was empirically parameterized without an explicit reference to the time-rapidity analogy, we found[197] that the formalism sketched already in our first paper on this subject[180] could explain the experimental observations and lead to a new scaling regularity (which we called beta-scaling). The superiority of this approach, as compared with other phenomenological approaches, lied in the fact that it permitted to relate the multiplicity distributions with Bose-Einstein correlations and led to new testable predictions, which confirmed our approach. Indeed, the increase with the width of the rapidity window of the normalized moments of the MDs in the rapidity plateau region could be parameterized by a power, and this led Bialas and Peschanski[198] to the suggestion that this behavior may reflect intermittency. Intermittency is a physical phenomenon observed in macroscopic physics, in particular in turbulence and as it names suggests it reflects periodic interruption of a certain process. It is due to the absence of a fixed scale in the problem. However soon after this proposal was published it was pointed out[199] that the quantum optical approach to MDs of the Marburg group, which implies a fixed scale, the correlation length, predicts a similar functional relationship between MDs and the width of the rapidity bins. One important detail had yet to be clarified: The quantum optical approach was applicable only to identical particles, while the initial experimental observations referred to charged particles, i.e. mixtures of positive and negative particles. And indeed, when identical particles data became available, it was found that the so-called intermittency effect was strongly enhanced, proving its quantum optical origin. This conclusion was confirmed by other independent studies.[200] From the analogy point of view the issue of intermittency in particle physics is an illustration of the limits up to which analogies are useful or applicable.

The above example refers to the plateau region of the rapidity distribution, but the importance of the quantum statistical considerations for MDs extends beyond that region. Thus the change of the form of the MDs with rapidity is under certain circumstances a signal of quark-gluon plasma formation[201] (cf. chapter 28) and also of long-range correlations.[202]

The importance of the quantum statistical approach to multiplicity distributions is reflected also in the fact that quantum statistics prescribes upper and lower bounds for the MDs. Indeed we saw that the two extremes of order and chaos mentioned above correspond to the Poisson and negative binomial distribution, respectively. However the Poisson distribution is the narrowest possible distribution, while the negative binomial distribution is the broadest. This led us to predict[203] that the experimentally observed broadening with energy of the multiplicity distribution in high-energy multiproduction processes must stop.

Self-Induced Transparency in Proton-Nucleus Reactions; A Filter for Production of Coherent Pion Beams?

Because of the finite extension of the rapidity plateau, one has to distinguish between phenomena, which depend on the position of the rapidity bin and those, which depend on its width. As mentioned above the story of multiplicity distributions in finite rapidity bins started in 1981 when we found[196] possible evidence for *self-induced transparency* in nucleon-nucleus reactions.

Optical self-induced transparency is the phenomenon discovered in 1967 by McCall and Hahn[204] in which an optical medium becomes transparent to a coherent pulse. If certain conditions are satisfied the front of the coherent pulse excites the scattering centers and the rear of this pulse de-excites them by producing stimulated emission.[205] This process repeats itself along the incident direction so that eventually the pulse will leave the medium without perturbing it: the medium becomes transparent to that pulse. In the above context, coherence means quantum optical coherence. As discussed in chapter 17 a different, although related kind of coherence, had been observed much earlier in cosmic rays proton-nucleus reactions. It manifested itself through the absence of "cascading": the amplification of the number of secondaries when comparing proton-proton with proton-nucleus reactions was quite small, not exceeding 2 for the heaviest targets. Landau and Belenkii interpreted this observation[206] as due to the formation of a "coherent" tube in nuclear matter. Subsequently the suppression of cascading was confirmed with accelerator data and on top of that a new remarkable property was reported: the increase of the multiplicity due to the nuclear target was concentrated in the backward hemisphere. Friedlander and collaborators named this phenomenon "transparency."[207] In 1981 in pursuit of our investigations on applications of quantum optical methods to high-energy physics, we asked ourselves whether there does not exist a deeper connection between optical self-induced transparency and the transparency due to the coherent tube observed in nuclear matter. More concretely, we asked whether the coherence of the nuclear tube is not of higher order, i.e. of quantum optical nature. This higher order coherence could have been created in the initial p-p encounter, in line with our previous observation of possible evidence for coherence in Bose-Einstein correlations. Then, if conditions analogous to those in optics are met, the nuclear medium acts as a filter for the incident pulse, allowing only the passage of its coherent component; a coherent pionic beam, analogous to a pionic laser would have been created.

A consequence of this hypothesis is that the hadronic beam, which manages to escape from the nuclear medium and which is seen in the backward rapidity region, has higher order coherence. This would mean that the multiplicity distribution in that region differs from that in the forward region: it should be narrower, more Poisson-like, while that in the forward region would contain an appreciable chaotic component. To our surprise, Erwin Friedlander found indeed, in emulsion data, possible evidence for just that effect. Encouraged by this empirical observation, Gerald, Erwin and I[196] developed a simple theory of the self-induced nuclear transparency in terms of an effective microscopic Hamiltonian for field sources of scattering centers (fermions) and fields (bosons) and showed that it admitted a soliton solution for the pulse that traversed the medium. This implied that the intensity of the pulse was not attenuated in the medium (cf. chapter 19), i.e. transparency due to higher order quantum coherence. A further not yet tested consequence of this effect is that Bose-Einstein correlations must follow the same pattern as that suggested by the multiplicity data; the BEC in the backward region should show more coherence than those in the forward region. Given its possible applications as a mechanism for the production of a coherent pion beam the experimental confirmation of self-induced transparency in nucleon-nucleus reactions is most challenging. It is still awaiting better proton-nucleus data. It is to be hoped that the relativistic heavy ion accelerator RHIC at Brookhaven will provide these data. The fact that one of the main collaborations of RHIC mentions this task in its recent White Paper[208] is encouraging and shows that the importance of this new phenomenon has not been forgotten.

Self-induced transparency in nucleon-nucleus reactions is an analogy, based on at least two other quantum optical analogies: higher order coherence and stimulated emission. To the same category belong two other quantum optical analogies, the phenomenon of superfluorescence, which we attempted to use to explain the Centauro effect and superradiance, which we applied for the interpretation of some special Bose-Einstein correlation observations in heavy-ion reactions.

Superradiance in Heavy Ion Reactions

Superradiance predicted by Dicke[209] is the phenomenon of cooperative photon emission by a large number of atoms, in analogy to the radiation of a classical macroscopic dipole. The radiation is coherent in the quantum optical sense and highly direction sensitive: for a pencil-shaped "cavity" the emission occurs mainly in the axial direction. The effect presupposes the existence of at least two "collective" levels, one of which is excited.

Inspired by some Bose-Einstein correlations measurements in low energy heavy ion reactions, which reported partial coherence, we suggested[210] that an analogous phenomenon might occur in heavy ion reactions. There were two experimental reasons that lead us to this suggestion. (1) If the radiation had not been due to a collective state, the coherence produced in the initial collision would have hardly survived final state interactions in nuclear reactions at these energies (1,8 AGeV). (2) The radiating fireball appeared to be non-spherical.

We developed a simple quantum mechanical model of a two level system made of nucleons and Δ resonances, which resembled a giant Δ resonance and we showed that under certain conditions such a system would behave as a macroscopic dipole emitting pions in a coherent state. Moreover and most importantly, if the source were not spherical the emission would be anisotropic, i.e. the coherence would depend on the emission angle.

Interestingly enough the experimentally observed degree of coherence depended indeed on the emission angle with respect to the longitudinal axis of the fireball. Furthermore, in support of our main assumption about a collective Δ resonance state, below the threshold for Δ production no coherence was observed. A prediction of our model was that the effect should also be seen in multiplicity distributions: the enhanced coherence should manifest itself in a narrowing of these distributions. A confirmation of this effect would be highly desirable.

Superfluorescence and the Centauro Effect

Superfluorescence is related to superradiance in the sense that it is also due to cooperative emission; if N is the number of sources, the intensity is proportional to N^2 and not to N. However, while in superradiance the emission comes from a system, which is in a "correlated" state, in superfluorescence the emission comes from uncorrelated individual sources (atoms in optics). In that case there is no initial macroscopic dipole; the effect is started by normal fluorescent emission, which develops into correlated emission, i.e. the "macroscopic dipole," not present from the beginning, is subsequently created and emits coherently.

Centauros are very rare cosmic rays events where there are no pions, while in standard events pions dominate. Another distinguishing feature is the much larger average transverse momentum of the particles. In Ref. 211 we put forward the hypothesis that these abnormal events are due to a collective fluctuation in the "normal" emission process. The bulk of normal events can be described by the Landau hydrodynamical model, according to which emission takes place only after the initially highly compressed and heated system expands and cools down to the characteristic hadronic freeze-out temperature T_f of the order of the pion mass. This explains why most secondaries are pions with an average transverse momentum $\sim 2T_f$. But *most* does not mean all. The exceptions to this rule are the few "leaked out" particles, which escape before freeze-out, i.e. at higher temperatures. From simple thermodynamical arguments follows that the masses of the leaked out particles are higher than those of the pions and this could explain the absence of pions and the higher average transverse momentum. Superfluorescence comes into this game through the fact that in this phenomenon fluctuations play a major role, inducing the decay of the unstable equilibrium state. A formalism similar to that used in predicting the effect of self-induced transparency does the job. A predicted consequence of this scenario is that, contrary to what had been assumed before Centauros do contain pions, but only of low energy.

While Centauros are still a disputed subject — some of the reported events have "disappeared" and there exist several alternative interpretations — the idea that superfluorescence could play a role in particle physics might be a useful one.

Squeezed States

These are generalized coherent states, which have also found applications both in Bose-Einstein correlations and multiplicity distributions. It was mentioned above that classical fields are coherent states. Coherent states are actually the nearest classical approximation of quantum states. This is so because they minimize the product of incertitudes in the Heisenberg indeterminacy relation, i.e. they allow for the maximum precision in the simultaneous measurement of momentum and coordinate, still compatible with quantum mechanics. There exist, however, also states for which the incertitude in one variable, say momentum, can be squeezed at the expense of the other, the coordinate, and vice-versa: these are squeezed states. The importance of this remarkable property lies among other things in the possibility to reduce quantum fluctuations, and this explains the great expectations associated with them in communication and measurement technology. It was probably this property that stirred my interest in squeezed states, although this happened, indirectly, through Apy Vourdas. Around 1986 Apy Vourdas, who had gotten his PhD in Manchester with a thesis on QCD and was a postdoc at Liverpool, had joined my Marburg group. Apy had also an electronic engineering degree and this, presumably, explained his simmering interest in communication technology on the one hand and quantum optics on the other. Actually, after leaving Marburg, Apy returned to his old love and has become one of the leading British experts in quantum computing. Since our Marburg group had been working for some time on the applications of quantum optical methods in high-energy physics, this "marriage" was quite a natural thing. It turned out to be also fruitful: among other things I personally became involved in squeezed states and Apy and I, besides a few joint papers on inelasticity distributions in high-energy reactions, published two papers on squeezed states, which had a certain impact and which will be commented below. On top of that, when Igor Andreev joined our group several years later, Igor and I found a new application of squeezed states, this time in the search for quark gluon plasma.

While ordinary coherent states are eigenstates of the one particle annihilation operator, squeezed states are eigenstates of the two-particle annihilation operator. Among others Bogoliubov used them for the description of electron-hole pairs in his theory of superconductivity.

When we began to work on squeezed states in the mid 1980s it was known that these states lead to certain specific features in Bose-Einstein correlations: while the intercept for coherent states is *1* and that for chaotic states 2, the intercept for squeezed states can take arbitrary values. Given the ongoing work of the Marburg group on multiplicity distributions, Apy and I started to study the effects of squeezed states in these distributions. We had then two surprises:

Surprise (A): Although the multiplicity distribution P(n) of squeezed states was known in the literature since the 1970s, nobody had apparently observed that it had a remarkable property: it was an oscillating function of n. This property could have been easily predicted from the fact that analytically P(n) was proportional to a Hermite polynomial H_n and Hermite polynomials are known to be oscillatory functions of n.

Surprise (B): For the more realistic case of superpositions of squeezed and chaotic (thermal) states we found that the oscillations disappeared, and that happened even for the smallest amount of thermal noise accessible to numerical estimate.

On top of these physics surprises, we had however a surprise of another nature: After discussing these amazing results with the other members of our group, Apy and I wrote a short note and on the day we wanted to send it for publication Fernando Navarra, who was at that time my PhD student, came to my office and showed me a paper by W. Schleich and J. A. Wheeler in the latest edition of *Nature*,[212] where they also had found our surprise A), i.e. the oscillatory behavior of the multiplicity distribution for pure squeezed states. Since they had not found surprise B), which actually was the true surprise, we sent off immediately our note, adding of course a quotation of Schleich's and Wheeler's paper. Our paper appeared as a "Rapid Communication" in *Physical Review*[213] and was followed by another one[214] where we applied these results to Bose-Einstein correlations and multiplicity distributions.

My next encounter with squeezed states happened a few years later when we found the particle-antiparticle correlations mentioned before.[187] That particle-antiparticle correlations are related to squeezed states can be understood by recalling that squeezed states are two-particle coherent states and that a particle-antiparticle pair is virtually equivalent with a pair of identical neutral particles. Another aspect of this relationship is the fact that the maximum of the intercept of the correlation function for two identical neutral pions, e.g. is *3*, instead of the canonical bunching value *2* for two identical charged pions; this "over-bunching" is a characteristic property of squeezed states.

But this is not the end of the story of the surprises due to squeezed states. In 1996 Igor Andreev and I came across[215] a new property and possible application of squeezed states. We found that squeezed states appear naturally when one deals with rapid phase transitions (explosions). Indeed, if one considers the transition from a phase a to a phase b and assumes that it proceeds rapidly enough so that the relation between the creation and annihilation operators and the corresponding fields in the two phases remains unchanged, it follows immediately that these operators obey the transformation properties specific for squeezed states. This property could be used, among other things, to determine whether the hadronization of quark-gluon plasma proceeds rapidly, through an explosion, as has been conjectured in some recent studies; the squeezed states would manifest themselves in the Bose-Einstein correlations, e.g. through overbunching.

The history of the application of quantum optical methods to multiparticle strong interaction phenomena has not been an avenue without obstacles. Some of the difficulties encountered were mentioned above. One reason for these difficulties was the belief that the specific features of strong interactions like final state interactions and conservation laws, which distinguish these interactions from electromagnetic ones, overrule any analogy with optics. However it was then realized that the effects of conservation laws are not important for high-energy multiparticle production phenomena.

As to final state interactions, reflected (in BEC) mostly through resonances, progress has been made in handling them[191,192] and the very fact that BEC of hadrons have strong similarities with those between photons proves that they cannot distort too much the quantum optical analogy.

Another reason why (some) practitioners in this field were reluctant to accept the lessons of quantum optics in particle and nuclear physics, and in particular the lessons which emerged from Glauber's work, is due to the fact that the quantum optical formalism is somehow outside the standard curriculum of subatomic physics and has to be learned before being able to use its applications. Actually, even in optics the impact of Glauber's pioneering work was apparently not appreciated by everybody. About three years ago I visited a US University that has a renowned center of modern optics and the colleague who showed me around was quite surprised to learn from me that I considered Glauber's contributions to quantum optics as fundamental. This explains in part why it took the physics community quite a long time to recognize Glauber's pioneering work in quantum optics by the Nobel Prize award. As to the applications of quantum optics to high-energy physics only three months before the award of the Nobel Prize, which my collaborators and I had expected for many years to happen,[a] I got an email from him in which, after telling me that he had been invited to the meeting of Quark Matter 2005 to give a talk, he wrote: "I still have no idea what to talk about. Have you any suggestions? I still have the impression you are the only person who has written about this field with any understanding — or even awareness — of what has been done in quantum optics." Fortunately it turned out that I was not the only person who understood what he has accomplished.

The story of the use of quantum optical methods in multiparticle productions is somehow similar to that of the Landau hydrodynamical model: hydrodynamics is also a topic, which is not usually taught in undergraduate courses. Besides that, and

[a]I had nominated him for this price twice.

most importantly, both approaches met resistance because they contradicted the preconceived lore: the Landau model contradicted the boost invariance assumption and the quantum optical analogy surpassed the wave function formalism. I personally (and Peter Carruthers as well) had the "privilege" to experience both manifestations of aversion.

24

Traveling to the East

When I visited the United States for the first time I underwent a slight shock of civilizations. This happened again in Asia, although the dissimilarities between Europe and Asia are quite different. Moreover, within Asia itself the disparities could hardly be bigger.

In India I was shocked by the strong contrasts between rich and poor. In front of elegant five star hotels there were tents in which people lived without water, electricity and all other amenities of modern life. At the same time the road was churned up by cable lying work replacing the old copper telephone cables by optical fibres. That happened in 1988, long before Europe or the US introduced this new system of communication. Going by taxi was another shocking experience. When the taxi stopped at a red traffic light, children extending their arms into the car immediately rounded it up — the windows were usually open because of the heat — and begged for money. That reminded me of Sao Paulo, Brazil: there the police recommended drivers simply not to stop at the red light, because of the danger of assault by armed gangs.

Japan was quite a different experience. In 1991 I spent there six weeks as an exchange professor and visited a number of universities. All over all it is the country which by its high degree of civilization impressed me most. Trains, among others the famous Shinkansen, functioned like Swiss watches. On the platforms of the railway stations there were marked stops for each coach so that with your seat reservation you knew exactly where to wait. On the train waitresses elegantly dressed in kimonos were serving meals. When entering and leaving the coach they greeted the passengers by deeply bowing. Japanese people mostly travel without luggage: There exists a perfectly organized city-to-city luggage sending system. When leaving the hotel you leave your luggage with the receptionist, indicating the address where you want to pick it up and usually it arrives there before you for a nominal fee. Actually, this is in part due to the fact that "business" hotels have very small rooms, without cupboards or wardrobes, but are equipped with kimonos and toothbrushes. The cleanliness in the cities is quite impressive. The taxi seats are covered by white linen wraps, which the driver changes after each passenger. The driver is dressed in an irreproachably clean uniform including white gloves. As to other cars, they look all brand new, as if in that country there would be no used cars. On the main streets of big cities service men are continuously polishing the brass plates of the banks and other institutions. This human effort in certain domains is matched by a high degree of automatization in other sectors of public life. In no other country, the US included, I saw so many sophisticated automats, where you could buy everything. The general technical level seemed to me superior to that of Europe and the USA. In the university guesthouse of the University of Matsumoto for example, and that happened in the early 1992s already, I had satellite television and could follow programs from all over the world. However, Japan was also the country where for the first time I felt a language barrier. Railway stations, bus stops and streets had their names written in Japanese only and you could hardly find anybody who would understand what you were looking for. Very few people on the street understood a minimum of English and even at the university there were hardly secretaries who spoke a European language. To find your way in town you had to have the name of the place you were looking for written in Japanese on a slip of paper and hope you will understand the signs of your Japanese interlocutor whom you may ask for help. Uli, who joined me two weeks later, learned this lesson on the bus.

She was returning from the city to the Matsumoto university campus but did not know at what stop to descend. Asking the driver did not help, the word university, common to all European languages, did not tell him anything. She recognized the entrance to the campus only after the bus had passed that stop. In many cities streets had no names and houses no numbers, so the coordinates of a house were indicated by words like "next to the school" or "across the pharmacy," etc. One of the main roles of the Japanese police is to find addresses of people. The Japanese who travelled abroad understand the difficulties a foreigner, who does not speak their language, when visiting their country. That is why Minoru Biyajima, the initiator of my Japan trip waited for me at Narita Airport where from we took the train to Tokyo and then changed to the train to Matsumoto, the first stop of my Japan visit. A few days later, my next host picked me up from there and the visit continued the same way.

In Matsumoto, a medium sized city in the neighborhood of the Japanese Alps, I had a strange experience. I had taken a bus to a spa called Utsukushigohara at the outskirts of the city and was strolling around, when I observed that I was followed by a dozen of children who entertained themselves quite vivaciously. When I turned around I realized that they looked at me with widely opened eyes and some of them approached and touched me, as if to convince themselves that I was not a ghost. I assume that they had not seen a white person with blond hair before, despite the presence, for forty years, of an important American military contingent, which is concentrated mainly in Okinawa.

We had another interesting experience in Tokyo, while visiting Namiki at Waseda University in Tokyo. Namiki gave a dinner party in our honor. Before that his secretary asked Uli whether there are things which we do not eat. Uli assured her that we ate everything, except perhaps raw fish. The outcome of this "understanding" was that we were served exclusively raw fish, albeit with different sauces. As to other eating habits, Japanese people sometimes express their appreciation of the food they eat, in particular when it is noodles, by loudly slurping. I assume, however, that this is a lower class phenomenon. I had another kind of shock in China, when I saw in a small village ten workers transporting a huge tree on ten bicycles. The contrast with Hong Kong, which I visited soon after that and where, on the main commerce streets, I felt myself rather on Oxford Street in London could hardly be bigger.

In 1991 I visited Petersburg, Moscow and Dubna. Half an hour before leaving for the airport I got an email message from my Moscow hosts who advised me to take with me some biscuits and dry sausage: in Moscow I was supposed to live in the guest house of the Lomonosov University and the cafeteria there was short of food. In the almost empty Lufthansa plane to Petersburg, my fist stop, the stewardess announced us that the plane will land at 17.20, local time. An hour later, after the captain had contacted the Petersburg airport, she corrected herself: the arrival time was 18.20. Actually we landed at 17.20. The confusion was due to the change from winter to summertime, which that year took place on another day than previously, but about which the traffic control of the airport was not correctly informed. Actually this was only the first sign of anarchy I would meet all along my visit. The airport of this city of 15 millions inhabitants was a shock. The very few buildings were actually barracks in a deplorable state. So were the roads along which the bus took us from the plane to the terminal. Shabelski and his wife waited for me at the airport and explained to me that being Easter Sunday there were no taxis available. Eventually we found for hard currency a "private" taxi. As a matter of fact, hard currency was the only currency with which one could buy something, and that, usually, only in the big hotels. The situation was better in Dubna, due to the existence of the Joint international institute, but my Russian colleagues there, and those in Moscow and Petersburg as well, complained about the inadequacy of their salaries, which expressed in Western currency, were absolutely ridiculous. I understand that now many of the aspects mentioned above have changed into the positive although the income discrepancy still persists, albeit to a lesser degree.[216]

Section IX

From Germany to France; Quark-Gluon Plasma; A Literary Intermezzo; Reflections (1995–Present)

25

Moving to Paris

In 1995, after having reached the age of 65 I had officially retired from the University of Marburg. This Hessen lawis quite strict, but it allows professors to continue to teach, to do research and to have PhD students, however without providing them with any financial support. In my case this lack of support went so far that the chairman of the physics department did not allow me to get financial support from the Ministry of Research or from the Deutsche Forschungsgemeinschaft for my collaborators anymore, in particular for Michael Plümer, who had sustained with brio his "habilitation" (the equivalent of a doctor of sciences degree, entitling him to get a professorship at a German university). The reason invoked was that Michael had already certain seniority and if he continued to be employed by the department, according to the German labor law he would automatically get tenure and the department would have to provide him with a position. A few years later I met on a social occasion the President of the University, who was at that time Werner Schal. When I told him the story he was astonished. He told me that this particular measure, about which he apparently heard for the first time, was quite unique and contradicted the interests of the university; indeed, through it the university had lost an appreciable amount of funds. As to the legal aspects of the issue, there were ways how to circumvent these. The department chairman happened to be the same person as the initiator of the report about heavy ion physics. Although some people saw in this a confirmation of the fact that the Einstein/Unsöld episode had not yet been forgotten and that the department was cutting in its own flesh, after twenty years of writing proposals and reports I was fed up and gave up fighting. I wanted to use the few years I still had left — my heart disease history was not encouraging — by working without constraints and administrative obligations.

For a phenomenologist the contact with people working in your field is essential. Since my group in Marburg was volatilizing I decided to move to a place where high-energy physics was actively pursued. Uli who had not yet reached retirement age, but also had serious health problems, decided to retire, as well. One possibility was to go to the States for a while. Given my long-standing collaborations, I would have had no problems to get a visiting professorship at an American university. However, for various reasons, among other things because our daughter Diana wanted to stay on in Marburg, where she felt quite at home, we did not want to move to a place too far from it. Two choices came up: London, where we had lived one year and which we had liked very much, or Paris. Paris was nearer to Marburg and easier to commute to. It attracted me also professionally and we had there many friends from Bucharest. It was also the only place in Europe where I still had some family, who had survived the holocaust. Last but not least, who would not have liked to live in Paris, if the possibility arose? So it was Paris, and so far, after twelve years spent there, we did not regret our choice. This move (Uli's mother joined us soon), turned out to be not only pleasant but also useful from the professional point of view; actually for a physicist these two kinds of points of view are hardly distinguishable. Financially this decision was an adventure, because we could not afford to buy a flat and had to rent. But somehow we succeeded to make the ends meet, albeit I had to place my working desk and computer in the bedroom.[a]

[a] Actually I am in quite good company: Nabokov wrote many of his novels in the bathroom.

Roberto Salmeron and Alfons Capella were quite helpful in arranging my association with the High Energy group of Orsay, the present Laboratoire de Physique Théorique (LPT) and from time to time I still participate in the seminars of this laboratory. At present, however, I attend more regularly, among other things because of the distance (it takes almost 90 minutes by public transportation from my home to Orsay), the seminars at Jussieu (Paris VI) and College de France. Driving in Paris and suburbs is at certain hours one of those things you better do without. This is also the reason I could not attend the frequent seminars at Saclay, which would have interested me; to reach Saclay would take me even more time than Orsay. Besides finishing the papers with my Marburg collaborators, which were in the pipeline, I started to digest my work on Bose-Einstein correlations, which extended over a period of twenty years. This process resulted in the publication in 1997 of the commented collection of reprints[193] on Bose-Einstein collections and in 2000 of a review[191] and of the textbook[192] on Bose-Einstein Correlations. In the mean time and in particular after 2000 I had time to think about more general aspects of physics, if not anything else, but about finding a common denominator for the dozens of quite different physical phenomena I had been involved since 1955; this led eventually to the present book.

26

France Versus Germany, Personal Impressions

French Impressions from a Scientist's Point of View

Paris is a wonderful place for somebody wanting to remain in touch with the newest developments in science. While Germany is a federal state in which the "Länder" have a heavy say in certain aspects of life, in particular in education, France is a centralized state, where most of the decisions, including schools and universities, are taken by the central government. Besides that France has elite schools of higher education like École Nationale d'Administration, École Normale Supérieure, École Polytechnique, École Centrale, while in Germany this kind of "elitism" was up to recently proscribed. Most of the French Elite schools are situated in Paris or its surroundings; this contributes to the intense scientific and cultural life of this capital.

Besides that, there are the 15 universities belonging to the Université de Paris, out of which Paris VI, VII and XI have important physics departments, and then there is the atomic research center of Saclay, near Paris. Last but not least, there exists an institution in Paris, which is quite unique in the world, the College de France founded in 1530 by the king Francois I, as a counterweight to the Sorbonne, with the explicit task of teaching three disciplines, not taught at that time at the Sorbonne: Hebrew, Greek and Mathematics. Access to the courses was free of charge and open to everybody. At present the spectrum of disciplines is much broader, including of course physics, but the openness of this magnificent institution has been maintained. Moreover, College de France is also a distinguished research institution, where one finds some of the most world famous French and non-French scientists. While at French universities professors must have a French PhD, College de France is also from this point of view an exception: here everybody who is scientifically qualified can teach. Anatole Abragam mentions in his autobiography[217] this peculiarity of the French university system that prevented him, with his PhD from Oxford and a distinguished scientific career, of getting a professorship at a French university. College de France embraced him and did not regret it.

I learned for the first time about College de France from Jacques Prentki in 1969 at CERN: He shared his CERN position in the theory division (he was at that time leader of the division) with a professorship at College. I was impressed by the fact that the only obligation College professors had was to give once a year a six weeks course, with a subject of their choice, at a level accessible to non-specialists. Unlike other universities, this course must not be repeated; the topic had to be changed every year.

Who are the "students" of College de France? These are not students in the usual sense, since there are no examinations and College does not grant scientific degrees. As far as I can judge the listeners in the physics courses are either students from the Parisian regular universities or physicists working in other fields. Quite a few are retired people, who want to keep in touch with recent developments; the courses, which are followed by seminars kept by invited specialists from all over the world, address the newest ongoing research problems.

The difference between the French and German higher education system is matched by the difference in the way research is organized in the two countries. In France most research at the universities is funded and organized by Centre National de la Recherche Scientifique CNRS, which in 2005 had almost 12,000 full time researchers, who had no teaching duties. In Germany most university research is in the charge of the universities themselves and professors and assistants have to teach. Research at German universities is funded in part by the Deutsche Forschungsgemeinschaft, the equivalent of the American National Science Foundation, and the Bundesministerium für Bildung und Forschung, the Federal Ministry of Culture and Research. This funding includes also salaries for young researchers participating in a research project lead by a professor.[a] Up to a point there exists a German equivalent of CNRS, the Max Planck Society, which however in 2005 had only 4300 researchers. Moreover unlike in France many of the Max Planck institutes are independent organizations, not associated with the universities.

For a physicist, and not only for a physicist, another striking difference between France and Germany is the role nuclear energy plays in the two countries. In 2004 France got 78% of its electric power from atomic plants, while Germany only 32%. Moreover, according to a decision taken some years ago, all nuclear plants in Germany will gradually have to close down and no new plants are to be built. In France, on the contrary, nuclear energy is thriving and the majority of the population seems to agree with this policy; the third generation of atomic power plants is now projected. In Germany not only the Green party, but also the social democratic party are strongly opposed to the use of nuclear energy. That France has a different policy on this matter has to do in part with the fact that France is an atomic power and the French appear to be very proud of that. This explains perhaps also why in France, in opposition to Germany, environmentalists play so far a minor role in French politics. This lack of interest in environment problems explains in part how the Government of Mitterand — a socialist — could fool the French claiming that the radioactive clouds stemming

[a]Most recently this constraint has been dropped: research proposal can be submitted and research groups can be lead also by scientists who are not professors.

from Chernobyl made a detour around France and, unlike Germany, Italy, Switzerland, Britain or Spain did not contaminate the French territory.

Linguistic Impressions

Another cultural policy difference is the place the national language plays in the two countries. Germans do not care too much about the increasing influence of English in day-by-day life — on the contrary, the young generation considers it quite "cool" to use English words in their conversation. A strange consequence of the strong influence of English in the German language is the fact that certain English words have been incorporated in the German language, albeit with a completely different meaning. For example the mobile, cellular telephone is called in German *handy*. Interestingly enough, this word has started to be taken over with its new meaning into the English language.

In France, however, the official policy at least, is quite conservative from this point of view; some might consider it even nationalistic. Not only does the government, in order to prove the importance of French in the world, encourage and support worldwide the "francophonie" — sometimes even countries like Bulgaria, e.g. are counted by French officials among francophone countries — but it also strongly opposes the use of English in international political meetings. Even the French media has ridiculed the recent walkout of Chirac, who speaks fluently English, from a European summit meeting in Brussels, because a countryman of his had dared to speak in English.

The fact that in Germany English is spoken by quite an appreciable fraction of the population, while in France this is much less the case is, of course, in great part due to the fact that French is a Latin language while English, like German, belongs to the West Germanic language family. However, it is also a consequence of the policy initiated by De Gaulle and continued by all French governments, both conservative and socialist, of cultivating the pride of the "grande nation."

Language is also an important factor in the integration of immigrants. From this point of view France is in a much better position than Germany, since most of its immigrants came from former French colonies, in particular from Algeria, Tunis and Maroc, where French had been the official language. In Germany, on the other hand, large proportions of the immigrant population, mostly of Turkish or Arab origin, do not speak German and live in language and culture "islands," without contact with the rest of the country. Still, this advantage could not prevent the recent riots in the suburbs of big French cities; such riots are (yet?) unknown in Germany. This difference may be due to the fact that street unrests are traditional in France. The recent nation-wide manifestations against the CPE law (cf. below), for an outsider quite a minor law, point in the same direction.

To finish this linguistic excursion, from at least one point of view the way the language problem is handled in France is much more reasonable than that practiced in Germany. I am referring to the orthography rules, which, in France, are established and controlled by the French Academy. Not so in Germany, where they have constituted a matter of heated controversy, and that not so much among linguists, teachers, writers or journalists, but among politicians. All ministers of education of the Länder, all Länder governments and the federal government as well were involved in the *Rechtschreibreform*. In the last ten years these rules have been changed three times, the school books and dictionaries had to be reprinted every time and the poor pupils had to go along with these changes. And at the end of this comedy writers, publishers and newspapers, which have not to abide by the official orthography rules — only official documents and school books have to — have announced that they will not go along with the changes introduced by the orthography reform and that each of them will use his own rules.

Law and Order

The passion for rules and the almost religious cult of *law and order* derived from it is for many people a specific German peculiarity. The Nazis have proved that this cult can degenerate: the *industrial* killing of millions of people is also a product of this pathology. Even if genocides have happened before, that the tools of Western civilization were used for this purpose is quite unique. Another illustration of this degeneracy is the strict bookkeeping in the Nazi extermination camps and the cynicism associated with the criminal acts of the regime. The Jewish community of Germany had to compensate the German state for the damages provoked during the 1938 *Kristallnacht* by the Nazi gangs, who carried out the orders of that same state; families of people executed in the camps had to pay fees for the execution.[218] The excesses of the law and order syndrome reflect a breakdown of common sense inherent to any normal human being. Those who executed them were able to abandon all principles of ethical behavior in favor of blind obedience, because they felt protected by "law."

Fortunately this syndrome, heritage of German history, seems to have lost its dangerous character and at present an outsider might even be amused by some of its manifestations. Thus the tendency to regulate everything has penetrated also the Deutsche Forschungsgemeinschaft (DFG) which has created a special institute dedicated to "Information and Quality Control of Research" (Institut für Forschungsinformation und Qualitätssicherung, IFQ), which, with its 25 collaborators, will study and develop methods to judge the scientific merits of the research projects funded by the DFG and of German science in general. It would be fair to mention that the attempt to measure quantitatively the value of a scientist, e.g. by his number of publications and the number of quotations of these publications, originated in the States. However to make of this doubtful fashion a "science" as the IFQ is obviously supposed to do, seems to me a typical German exaggeration.

Last but not least, the case to be quoted below might sound like a joke, but it is not; it happened indeed. In the Land of Baden–Würtemberg there exists an old folk custom, the cowpat roulette. The game takes place on a lawn divided like a chessboard into squares and it consists in betting on which square a cow will place its first pat. Those who guess correctly, win. The trouble started when an animal protectionist complained to the Ministry of Agriculture of the Land that the cowpat roulette might infringe the dignity of the animal. The view of the ministry that the "sentiments of a cow which is watched and occasionally spurred on by a crowd of people while depositing its pat, cannot, at present, be checked scientifically" did not convince everybody and produced an intense cow-ethics debate.[219]

In a certain sense the French suffer from an opposite syndrome: they do not take very seriously rules and even laws. While the typical French nonchalance, which is part of this behavior, is up to a point a pleasant aspect, its exaggerations are sometimes less amusing. This refers for example to the fact that you can see men urinating along the streets in the middle of Paris. It also refers to the disrespect of traffic regulations, including driving after consuming alcohol beyond the legal limits. But it also manifests in the frequent strikes — France is the world champion of strikes — in the blocking of highways and railways, in the burning of harvests. After renting a flat in Paris the janitor introduced us to the "rules" of the apartment house, telling us, from the beginning, which rules are respected and which not.

Social and Political Differences

Although France and Germany belong to the founders of the European Union and the economical and political relations between these two countries are already for half a century quite close, there are still remarkable differences between these two countries. One of these refers to the relations between employees and employers. In Germany these relations are ruled by laws, which make imperative negotiations between the two sides, represented by the employers organizations and the trade unions, respectively. This applies in particular to changes in the labor relations laws, which have become necessary because of the strong increase of unemployment. While in Germany these reforms where adopted more or less smoothly, in France the situation is completely different. A recent illustration of this fact is represented by the events related to the CPE — *Contracte Première Embauche* (first employment contract).

In an attempt to reduce the 25% unemployment among the youth, in spring 2006 the government initiated the CPE law allowing for a liberalization of the labor-employer relations. The French Parliament, where the center-right government lead by Villepin had a comfortable majority, approved the law, but even before Jacques Chirac, the President, had a chance to promulgate it, under the pressure of the street he had to water it down, almost annihilating its main reformist content. Still, this retreat was not sufficient. Manifestations against the law organized by the trade unions and university and high school student organizations all over France started to paralyze the functioning of the country: the majority of universities were blocked by protesting students, who prevented those willing to attend courses to enter. Many high schools were in the same situation. On top of that the center of Paris was invaded by "professional" rioters, who came from the underprivileged suburbs of Paris, and who repeated here the havoc they had produced in their neighborhoods in autumn 2005: destruction of shops, cars, even schools, and battling with the police. They occupied the Sorbonne where they destroyed computers, libraries, and furniture. Similar actions took place in the province. These rioters had nothing to do with the universities or with the labor unions. On the contrary, the CPE law was intended to help the "underprivileged" finding a job. The main reason for this violent reaction was that the trade unions had not been consulted before the law was formulated. Although labor unions had obviously no right to interfere with the decisions of the Parliament, they saw in this an infringement of their authority and role. On the other hand if one considers that on the average less than 10% of the working population of France are enrolled in trade unions and that there are five different and competing trade unions one wonders where the disrupting power of these organizations comes from. The answer lies in part in the fact that this small percentage of trade unionists comprises only 5% of the total number of workers in the private sector but 30% of the total number of workers of the public sector (state employees and civil servants). And an appreciable fraction of the French economy is still state owned, that is belongs to the public sector. Besides that, among the civil servants ("fonctionnaires") the teachers constitute a very influential group, which votes traditionally with the left. Quite a few are still communists and even Trotskists and for them the center-right parties represent the class enemies. In the CPE dispute in many cases teachers encouraged their pupils to strike.

For somebody not acquainted with the French society other facts are puzzling as well:

— The important role university and high school students play in the political and social life.
— In many, if not in the majority of universities and high schools, voting for strikes is done in public meetings by raising the hands and not by secret balloting. The democratic character of these ballots is highly doubtful. The similarity with habits in communist and fascist countries is striking and hardly accidental: In no other Western country (with the exception of Italy) the influence of the communist party on public life had been stronger, despite the fact that in recent years the number of votes this party got in elections has shrunk from 30% to less than 5%.
— The important role student organizations have in *high schools*. Is this also a sequel of the communist influence? (In the former Soviet Union and its satellites student organizations and trade unions had the role to "transmit" the party line to the "masses"). Anyway, I am not aware of any other country where this happens

at present. Actually, this role of student organizations in high schools is hard to reconcile with the fact that in France, like in most other countries, the right to vote is reserved to people who have reached the age of 18.

— The majority of the French population was apparently opposed to the CPE law, although most economists believed that this law might have had a positive role in reducing the unemployment.
— Strikes *precede* negotiations between employees and employers and do not *follow* in case these fail.
— The French government has apparently ignored this mood among the population; had it proceeded with more diplomatic and pedagogical skill, it might have got its CPE project through, albeit not in the short delay it wanted.

The 2006 events remind one about the rejection in 2005 by the French of the European constitution, despite the fact that most people should have been aware of the immense benefits and advantages the European Union had brought their country, perhaps more than to any other member state. Some commentators see in this a more general, European phenomenon: In opposition to the 1968 generation, today's generation is the generation of "no, no," no reform, no change. This reflects a lack of communication between politicians and people; politicians have failed to explain their voters the reasons and meaning of the new laws, the present situation of France within Europe, and that of Europe in the rest of the world. On the other hand the CPE incident is also the newest illustration of a fact observed in France since the 1968 events: sometimes France is not governed by its democratic institutions, but by the street. This development is not so surprising, after all: in Grasse e.g., a Southern France city renowned for its perfume industry, at the cultural center teenagers have been taught what in French is called *revendication* (claim). They were instructed how to formulate slogans and how to prepare banderols. Some people saw in this the raising of a generation of protesters and claimants.[220] Actually, in the protest meetings of 2005–2006 also participated youngsters aged 12–14. What is much more disturbing is that the disrespect of laws is not limited to the behavior of the man on the street, but extends to the very justice system supposed to guarantee the correct application of laws. This is reflected among other things in the abuses of power of examining magistrates and the way the principle of presumed innocence is handled by them. The now world famous Outreau case has made recently headlines, but before discussing this case I will mention another incident in which I was personally involved, and which I believe could not have happened so easily in Germany.

The Principle of Presumed Innocence

Both France and Germany are democratic states governed by the rule of law. This has important implications also for the justice system and is one of the main achievements of modern civilization. Somebody like myself, who spent a great part of his life under dictatorships, particularly appreciates the privilege of living in a democracy. That is why I was shocked when soon after moving to Paris I experienced here an outright violation of my citizen rights. I probably would have forgotten this incident hadn't it almost repeated itself quite recently.

In 1997, returning to Paris after a few weeks absence we found in our mail box a note from a bailiff announcing us that he had seized our joint Paris bank account on the name of *Monsieur et Madame Richard Weiner*, because the wife of Richard Weiner, Annick Weiner has an unpaid debt of 100,000 Francs. Besides the fact that we had no unpaid debt whatsoever, either in France or in any other part of the world, Uli's first name is not Annick but Nina Ulica. The bank, of course, not only knew Uli's correct name, but also our birth-dates and places of birth; at the opening of an account, we had of course to present our passports. This rule applies both to French and foreign citizens. Our birth dates and birthplaces certainly did not coincide with those of Richard and Annick Weiner sought by the bailiff. Not only didn't the bank and the bailiff check these details, but also, and that is much more amazing and alarming, the judge who had to approve the seizure of the account, did not either. According to

French (and other democratic countries') laws, the decision to seize a bank account can only be taken by a judge. The incident was closed and our account reopened after presenting our IDs to the bailiff. The fact that we were foreign citizens and that I warned him that I might get the German embassy involved, could have contributed to this swift action, which was followed by letters of apology from the bailiff and the bank.

Didier GATIMEL Patrick OKERMAN
HUISSIERS DE JUSTICE ASSOCIES
SIRET 397786104 00018 CCP PARIS 4.237.17.T
TEL 01 47 66 87 52
3 Rue de Logelbach 75847 PARIS CEDEX 17 FAX 01 47 66 82 61

PARIS ,le 09/04/97

REFERENCES A RAPPELER IMPERATIVEMENT
Dossier N°:6050858
DEFOSSEZ CLAIRE
C/ WEINER ANNICK

VOS REFERENCES

Mr RICHARD WEINER
177 RUE DE LOURMEL

75015 PARIS

—

Monsieur,

les renseignements bancaires qui m'ont été communiqués étaient manifestement erronés, je vous remercie de bien vouloir excuser cette regrettable erreur.

Enfin, je n'ai évidemment pas manqué de procéder à la mainlevée de votre compte bancaire le huit avril 1997, vous trouverez sous ce pli copie de cet acte dûment régularisé annéxé à la présente.

Recevez, Monsieur WEINER, l'expression de mes salutations distinguées.

However neither the bailiff nor the bank appeared to be embarrassed, a possible indication that this incident was not very unusual. We had quite recently (in 2006) the opportunity to test this conjecture, when I got from *Intrum Justitia*, a company specialized in collecting unpaid bills a summons to pay an insurance premium of 1000 Euros to an insurance company we never heard of. The summons was apparently a repetition of previous ones, which we never got, presumably because it was sent to another address, the address of the homonym. Although I pointed out immediately that it was an obvious mistake and that I have never been insured at that company, I got several more and more menacing summonses and eventually a letter from a (new) bailiff threatening with court action. Annoyed by this story I eventually charged a lawyer who got the record straight. To decrease the probability of repetitions of such incidents I took out our names from the Paris telephone directory, since I suppose that this was the source of confusion.

I wouldn't have mentioned the above incidents, which up to a point have a comedy character, hadn't in the mean time something really serious and tragic become known, which proves how far the non-respect of the principle of presumed innocence by the French justice can lead. I am referring to the Outreau affair, a justice scandal, which put under question mark the entire French justice system.

The Outreau Affair

In 1995–2000 eighteen people of Outreau, a poor suburb of Boulogne in Northern France, were accused of rape and other attacks on children. One couple admitted abusing their four children and were jailed for 20 and 15 years respectively. Their two neighbors received sentences of six and four years for lesser offences. However several of the other fourteen, who all pleaded not guilty, waited in prison for four years (and one committed suicide there in the mean time) before being acquitted in 2005 by a Paris appeals court after it became clear that they were innocent. Most of the people involved were parents, and their children were separated from them during all this time. During the appeal, psychologists and social workers admitted that their original findings were flawed. Two children admitted they had made up their stories. The professional and family life of those freed, including in particular their relations with their children, were irreversibly destroyed.

Besides the much exaggerated power a single examining judge has in sending to and keeping in jail a defendant, at the root of this almost unique affair is the flawed French justice system, where the principles of *habeas corpus* and presumed innocence, the corner stones of a modern democratic society, are arbitrarily handled and mishandled. Actually, in an extraordinary step, the prosecutor was forced to describe in the appeals court the affair as a "disaster" for French justice. He apologized to the defendants and declared: "We must make sure this never happens again." The President, the Prime Minister, the minister of justice presented official excuses to the victims in the name of the government and of the judiciary institutions. Furthermore, to help prevent repetition of such judiciary miscarriages changes in France's penal code are envisaged. For this purpose the parliament instituted a special enquiry, which was broadcasted live on television. The hearing of acquitted persons by the parliamentary enquiry caused a great emotion through the whole country. The magnitude of this affair reminds one of the Dreyfus affair, another justice scandal, which happened in France more than hundred years ago and which had important repercussions for the entire French society.[b]

Foreign Policy

Despite the close links between Germany and France, which have characterized their Post-War relations and decisively influenced the

---

[b]Although both these events were made possible because of a malfunctioning justice system, the Dreyfus affair had also another, political aspect: it was the result of a conspiracy, which reflected and was in part a consequence of the prevailing anti-Semitism in the French society of that time.

face of whole Western Europe, and despite of the increasing role of the European Union in the policies of its member states, there are still significant differences in the foreign policies of Germany and France. This refers among other things to the relations with Africa and the USA, and to the Palestinian conflict.

As an old colonial power France has tried to cultivate its relations with the African continent — the francophonie is part of this policy — although the success could hardly justify the means invested. Germany had already lost its colonies in 1918 and the fact that it was defeated in both world wars discouraged any significant attempt in this direction. While up to the Iraq war Germany had been, besides Britain, the closest European ally of the United States, the relations between the USA and France were much cooler. This went so far that De Gaulle withdrew France's defense forces from NATO, threw out non-French NATO troops and obligated NATO to move its headquarters from France. Only in 1993 France rejoined the NATO military command. This post-war French anti-Americanism has two roots. A) The frustration following the realization by the French establishment and De Gaulle in particular, that France had lost in the 20th century what it had been in the previous five centuries: the center of political initiative in the world and the status of big power. B) The influence of the communist party, which was a faithful executer of the policy of the Soviet Union.

In the Palestinian conflict Germany has been a traditional supporter of Israel, while since 1967 France became the most pro-Arab country of Western Europe.

Anti-Semitism

The pro-Arab foreign policy of France had also implications for its internal policy. This refers first of all to the coexistence of the Jewish and Arab communities in this country; France has not only the highest number of Muslims in Europe, but with the exception of Russia, also the highest number of Jews (around 600,000). But it refers also to the more general problem of Anti-Semitism. The Pro-Arab orientation of France was inaugurated by De Gaulle in 1967, after the Six day war. The fact that he called, with that occasion, the Jews an "elitist people, sure of itself and dominating," did not facilitate the coexistence between Arabs and Jews, either, nor did it contribute to the weakening of the "traditional" Anti-Semitism in the French society, which had manifested itself quite virulently, for instance, in the Dreyfus affair. Indeed, in a nation like the French, which considered itself in the not so remote past "la grande nation," such a statement almost unavoidably produces resentments. It reminds me of an anecdote about a psychiatrist, who presents to his colleagues his most interesting cases. The first one believes himself to be Plato, but otherwise behaves completely normal. The second one also behaves normally, except that he thinks he is Leonardo da Vinci. The third one, however, according to the host, is a really pathological case: he claims to be Napoleon. After an extensive examination of the three cases the visitors ask: "What is so special about this Napoleon, he also seems quite *normal*." "That's obvious" was the reply, "Napoleon that's me."

In France, like in almost all other European countries, the classical Anti-Semitism has its roots in the traditional anti-Jewish attitude of the church, be it catholic, orthodox or protestant. Initially the Jewish faith was for the church less a competitor than an ideological challenge. Indeed, unlike the Christian or Muslim faith, the Jewish faith does not proselytize. The proselytism of the Christian church explains why the Christian Anti-Semitism was not racist. Actually in the middle Age Jews could often survive by converting to Christianity. This was not possible in Nazi Germany or in most countries under Nazi rule: During World War II, the majority of European Jews, including those converted, were exterminated. Still, the anti-Semitism practiced by the church and its royal representatives in the middle age and that practiced by Hitler in the 20th century have something in common: the economical interests, the greed for the fortunes and goods possessed by the Jews.

The holocaust, as well as the creation of the state of Israel brought about a change in the intensity and in the nature of European Anti-Semitism. Besides the almost physical disappearance of its object, the weakening of the Anti-Semitism in Europe is mostly due to the fact that under the pressure of the victorious Allies Anti-Semitism and racism in general were explicitly declared a crime. Another factor contributing to this was the change of the attitude of the Catholic Church, in great part also due to the pressure of Allies, and of the United States in particular. Under successive popes the Catholic church officially apologized for the crimes committed by it against Jews, the "elder brother" of Christians, during its two-thousand year history. In France an obvious symptom of this change is the striking fact that a Jew converted to Christianity, Jean-Marie Lustiger, whose family perished in Auschwitz, was for more than twenty years archbishop of Paris and, as cardinal, was even among the possible successors of Pope John Paul II.

The pro-Arab bias of the French foreign policy acted effectively as a new source of Anti-Semitism in France: it became "politically correct" to condemn the policy of Israel, the more so that some actions of this state were not easy to defend even by its sympathizers. While until 1967 Israel was considered the "underdog," after the Six Days War the Palestinian Arabs took over this role. An appreciable proportion, if not the majority of Jews, although not Zionists — the fact that they live outside Israel proves that — identified themselves, or were identified by the surrounding society, with the cause of Israel and therefore from an Anti-Israel attitude to an anti-Jewish one was only a small, and more often than not, unperceived step. This step is particularly easy if the history of the creation of the state of Israel and the role of the Arab countries in the Palestinian conflict are ignored. Actually, the link between Jews and Israel is, besides its historical aspect, of more pragmatic nature: since its creation Israel has manifested itself as a strong defender of the rights of Jews all over the world. Moreover, many Jews, particularly in Europe, consider Israel as a possible haven in case history might repeat itself.

This new kind of Anti-Semitism is clearly seen in the anti-Jewish acts of violence in the underprivileged suburbs of big cities in

France, with their dominant Muslim population of Arab descent, but where also an appreciable number of Jews live. That this phenomenon is — at least on a big scale — not seen in Germany is probably explained by the fact that the number of Jews who presently live in Germany is much smaller than that in France. Furthermore, the denazification imposed by the World War II allies in Western Germany, however imperfect it was, bore some fruits. Even if, initially, not out of conviction, but rather out of opportunism, the West German society has proscribed Anti-Semitism and racism in general. The comparison between Western and Eastern Germany is in this sense quite illuminating. The former German Democratic Republic under communist rule considered itself, by definition, immune to Anti-Semitism. In reality the rigid communist dictatorship prevented not only Anti-Semitic and racist expressions of opinion and manifestations, but expressions of opinion and manifestations in general. Furthermore, there, as in the entire Soviet bloc Zionism was proscribed as a nationalist manifestation of capitalism and after 1967 almost all communist countries severed their relations with Israel and, unlike West Germany, adopted a net pro-Arab attitude in the Palestinian conflict. The consequences of these policy differences between East and West Germany have been seen after reunification. The freedom brought by this act to East Germany was sometimes interpreted in the "New Lands" as freedom to persecute and even kill people of other color, race and faith. This lead to the establishment in East Germany of "no-go areas" for colored people.

Up to a given point, a similar phenomenon has been seen in France: France, too, lived, until recently, under a double illusion: that it had been a country of resistance and it had therefore nothing to do with the holocaust, that the deportation of 75,000 Jews from France was a pure German affair. Actually, neither was the resistance in France a general phenomenon, nor was the deportation of the Jews possible without the active collaboration of the French authorities. This delusion cultivated under most French presidents, including Mitterand, who until a certain moment had personally collaborated with the Vichy regime, was officially exposed only in 1995 by President Chirac, who acknowledged the role the French state, represented by the Vichy government, played in the deportation of Jews.

The Anti-Semitism in Muslim dominated suburbs in France has also consequences for the non-Jewish population: in many schools of these suburbs the teachers are prevented by students to talk about the holocaust. And this although in France, like in Germany, the negation of the holocaust is a crime.

Political Correctness in Germany and France

When comparing Germany with France one has to note that with respect to Israel political correctness has in France a different meaning than in Germany: Post-War Germany has led, in general, a rather pro-Israeli policy, although it has often criticized the actions of the Sharon government and has consistently tried to cultivate its relations with Arab countries. A recent incident, which represents an exception in this specific German political correctness, is the scandal around the liberal party, Freie Demokratische Partei (FDP). One of its leaders, Möllemann justified terrorist attacks against the Israeli population by the policy of the Israeli government.

Möllemann's motives and acts were anything else but simple reflections of moral or political convictions, as their author tried to present them. It happened that Möllemann, a former vice-chancellor and minister in the German government, as well as vice-president of the liberal party FDP, was also president of the German-Arab Society. And in 2003 it was officially proven, that his anti-Israel campaign had been funded by Arab circles from outside Germany. Because of this and other wrongdoings Möllemann was indicted and faced many years of prison. Strangely enough, it took the FDP quite a long time to part with Möllemann, who eventually committed suicide.

The Möllemann affair was hardly over when in autumn 2003 a new anti-Semite scandal upset the political scene of Germany, the Hohmann affair. This time it involved a leading member of the Christian democrat party, CDU, one of the two major parties of post-war Germany, and this time his party, after initial hesitations, condemned quite soon his actions.

Of course, the way official Post-War Germany has handled the issue of Anti-Semitism has been dictated by its past. However, the desire of closing definitively the chapter of its Nazi-past has also led to a new aspect of political correctness, determined by the way Germany's actions are judged by the outside world. A frequent argument in the attempt of German politicians to fight against attacks on foreigners, anti-Semitism and other rightist manifestations is to invoke the negative reaction, which these acts would produce abroad. Without intending to doubt the good intentions behind this kind of arguments, especially if they come from high ranking politicians like Roman Herzog,[221] at that time the president elect of Germany, one sometimes has the impression that it is not so much the wrongdoing in itself that the authors of these statements condemn, but its effect outside Germany. This reminds one of the situation after World War II, when Germany desperately tried to become again an accepted member of the society of nations, out of which it had catapulted itself through its unprecedented crimes against humanity. Being aware of the fact that the German people had not yet really understood what it had done, but wishing to obtain immediate results, some of its politicians renounced to invoke questions of moral principles and preferred to argue that "good behavior" is the only way to get a re-entry ticket into the civilized society. As a matter of fact, child education sometimes proceeds this way: if you don't have time or patience to tell your child *why* something he did is wrong (or if you don't know yourself why...), you just tell him "this is wrong" and enforce this statement by sheer power.

Whatever progress has been achieved in fighting Anti-Semitism one must not overlook the disturbing fact that synagogues and Jewish institutions in big cities of Germany after almost six decades of democracy are still protected by police. If this is a reflection of a genuine danger, then it is an unsettling phenomenon indeed, which is not made less disturbing by the fact that this happens in France, too.

Another specific aspect of political correctness in Germany is the attitude towards the United States of America. Before the Iraq war and in particular during the cold war Western Germany's policy was strongly pro-American. That was dictated in part by the realization that only the alliance with the United States and NATO could guarantee the security of the Federal Republic. This policy was initially opposed for reasons of principle, but sometimes also for less moral grounds, by the left and by the "peace movement," but eventually it became, together with the reconciliation with France, *the* cornerstone of Germany's foreign policy. With the fall of the wall between East and West Germany and the reunification this attitude slightly began to change: Germany gained more self-confidence and relied more on its ties with the European Union, trying at the same time to improve its relations with Russia. The Iraq war constituted for Germany a watershed in the relations with the USA. The decision by chancellor Schröder to say no to any intervention in Iraq, *before and whatever the outcome of the debate within the Security Council of the United Nations would be*, lead to a massive deterioration of Germany's relation with the US. If during the cold war the alliance with America constituted the essence of what was considered in Western Germany as politically correct, the situation changed completely with the Iraq war. The old anti-Americanism of the peace movement was again "in". This change went so far that during the Iraq War, it was not Saddam Hussein that was the target of the attacks and of the fury of the street demonstrators, but president Bush, whom the Minister of Justice of Germany compared with Hitler.

Schools and Religion

A common school problem in France and in Germany due to Islamic fundamentalism is the clothing of Muslim female students and teachers. The growing Islamic fundamentalism manifests itself also in the way Muslim women dress themselves: the veil, which is standard in Muslim countries, has been imported to Western Europe and attempts have been made by some Muslim parents to impose it to their children in schools. As to adult Muslimwomen who exercise in Europe professions, which they could not dream of in Islamic countries, many of them wear the veil as a manifestation of independence, if not superiority. This "fashion" has produced a lot of trouble, particularly in a country like France, which is by constitution a secular state, but also in other countries like Germany, where the separation between religion and state is not so explicit. In schools many non-Muslim parents saw in this tendency an attempt of Islamic proselytism, in particular when it was practiced by Muslim teachers. The French parliament voted a special law forbidding the veil and other religious symbols in schools, both for students and for teachers, and many German lands did the same for civil servants, which include teachers. However, students in German schools are allowed to wear veils. For some Muslim students this measure did not go far enough: Most recently two Islamic students appeared in a high school in Bonn in burqas, which cover the wearer's entire face except for a small region around the eyes.

Schools in Germany have to fight with another problem related to the Muslim immigration and which does not exist in France: in some schools in big cities the vast majority of pupils come from families where one does not speak German. Under these conditions teaching in schools has become next to impossible for German students, as well. The case of the Rütli School in Berlin, the director of which had asked the Senate to close it down, has made recently headlines.

Another difference between Germany and France related to religion is the way the separation between church and state is implemented in these two countries. Thus in Germany (but not in

France) churches and religious communities can get the right to levy church taxes (*Kirchensteuer*) on their members and in most cases the revenue is collected by the state. The famous German writer Max Nordau in his book "Conventional lies" had already in 1883, fiercely criticized that concession towards churches, but apparently with no avail. Furthermore, while in France religion is not taught in (state) schools, in Germany it is, and students up to the age of 14 have the choice between catholic and protestant religion. After 14 in most lands they can chose also another discipline, instead, ethics or philosophy. It is worth reminding that the secularism ("laïcité") France is rightly so proud of and which, with the exception of the United States, is quite unique among democracies was introduced in the French constitution in 1905, as a lesson drawn directly from the Dreyfus affair. After the 1789 revolution this is probably the most important political contribution France has brought to modern society. It is the natural extension of the principle of freedom of choice and practice of religion to freedom of thought in general. For Voltaire, one of the spiritual fathers of the French Revolution, God was an unnecessary hypothesis. Although it might sound presumptuous, I myself agree with Voltaire on this point. That is why I also consider the anthropic principle propagated by some physicists and which claims that the values of constants of nature have the particular values we observe in nature in order to make life and our existence possible, non-scientific and unacceptable. It represents a step backwards and a negation of the Copernic principle, according to which the position of human beings in the cosmological order is in no way privileged.

German–French Convergence

To conclude this comparison between political and social aspects of France and Germany, it seems that, despite still existing differences, the two countries rapidly converge. This is not surprising after all, if one considers that the French president and the German chancellor, accompanied by many ministers, meet several times a year for an exchange of experience. The approach between Germany and France went so far that in one of these last meetings the two sides proposed that all citizens of one country should be entitled to get, upon simple request, the citizenship of the other country.

I could personally follow some of these convergence developments initiated from "above" and I will mention here two examples:

(a) The German and French health policies of the last ten years. To start with an amusing detail: both France and Germany considered (and continue to do so) for decades their own health care system as the best in the world. What is however a fact and a quite essential one is an important recent change, this time at the European Union level and of which we, as commuters between France and Germany, have profited in particular: citizens of the European Union can use medical services, covered by their national health insurances, all over the Union.

(b) History in these two countries will be taught in high school from a common textbook. Interestingly enough, it took a commission several years of work to edit this book, not because the views on World War II were difficult to reconcile, but rather the opinions on the role of the USA in the world after 1945; the German historians considered the French being Anti-American, while the French historians considered the Germans *atlanticists*.

The most important outcome of the European unification process and of the close relationship between France and Germany, in particular, has been of course, the fact that for the last sixty years there was no major war in Europe. It is a great feeling to pass the border between Germany and France, which constituted within only 20 years of the last century the scene of two devastating wars, not only without being stopped by a custom officer, but without even seeing such an officer; the customs buildings are deserted. To quote Alfred Grosser: the most convincing sign of the fundamental change in the European landscape is that our children, when confronted by their parents with this fact, have nothing to say but: so what? This statement is even more significant, if one considers that it is made by somebody, who was born in Frankfurt and who in 1933 took refuge, with his Jewish parents, in France, and who after the war dedicated his entire activity to the eradication of barriers between the two countries.

The closeness between France and Germany on official levels is not only a consequence of the political will of the respective governments, but is often a consequence of social movements which propagate from one country to another. The starting point of these movements was almost always France: The 1789 revolution, the secularism initiated by the Dreyfus affair and the 1968 student movements are typical examples. An exception may have been the aborted CPE law: it was inspired by recent German legislation under the social democrat government of Gerhard Schröder, and the fact that it did not materialize on the French side confirms that the wind in Europe usually blows from West to East. Actually the defeat of the Villepin government already seems to find its resonance in Germany: So far tuition in German universities was free, but recently some lands introduced fees and this lead to strong protests by the students. In Hessen e.g. at the end of May 2006 the students blocked highways and railways and in Marburg they occupied for several days the administration building and the office of the President. And what is quite symbolic, this movement is done under the French tricolor, and the CPE is quoted in the students meetings as a message addressed by France to Germany.

This not withstanding, it would be a great pity if this convergence, accelerated also by the world globalization, would make the differences in certain traditions and culture, as well as certain specific amenities disappear. Some of these differences constitute a valuable heritage of European history and should be preserved. To these amenities belong also the 300 kinds of French cheese and the inimitable French baguette freshly baked every two hours, all days of the year, Sundays and holidays included. But the 30 kinds of excellent *dark* bread, found only in Germany, must not be forgotten, either.

Section X

Quark-Gluon Plasma; A Literary Intermezzo; Reflections

27

Quark-Gluon Plasma, Another Old Love

Studying "life" the analogies between Germany and France fortunately did not prevent me cultivating my old hobby, the analogies in physics. At the end of 2004 K.K. Phua asked me to write a review on a subject of my own choice for the International Journal of Modern Physics. I accepted this invitation among other things because I could combine it with the writing of the present book. I chose as subject Quark-Gluon Plasma (the actual title of the paper[222] became "Surprises from the Search for Quark-Gluon Plasma? When was Quark-Gluon Plasma seen?"), because at that time the results of the new Brookhaven relativistic heavy ion accelerator RHIC had become known and this review gave me the opportunity to show that in terms of the previous Marburg work some of these results, which appeared to be quite surprising, were not so surprising, after all. Furthermore contrary to the dominant opinion, according to which heavy ion reactions are a necessary pre-requisite of finding the Quark-Gluon Plasma (QGP), I attempted to point out that this new state of matter might have been discovered already in particle reactions. In the following I will resume the facts, which lead to this conclusion. For more details the reader is referred to the article quoted.[222]

The search for quark-gluon plasma, sometimes also called quark matter, has constituted for the last three decades one of the highlights of high-energy physics. The great interest in this topic is easy to explain: Quark-gluon plasma is a new state of matter made of quarks and gluons, which is not found in everyday terrestrial environment. However it played an essential role in the evolution of the universe about 10^{-6} seconds after the Big Bang, when it was *the* phase the universe was in. In analogy with other forms of plasma found in nature, made of electrons and ions and in which the *electric* charge is screened, in QGP it is the "strong" charge color, which is screened. Its existence was conjectured in 1975 by Collins and Perry[223] (the name QGP was actually coined in 1978 by Shuryak[224]) after the establishment of quantum chromodynamics (QCD) as the theory of strong interactions. QGP as a state of matter is a consequence of *asymptotic freedom*, one of the most salient features of QCD. It consists in the fact that while at large distances, that is outside hadrons, quarks and gluons interact strongly and are therefore *confined*, at small distances they become free. The discovery of QGP in the laboratory would constitute a brilliant confirmation of QCD in general and of asymptotic freedom in particular.

Where, Actually, Does Asymptotic Freedom Begin? From Supefluidity to Strongly Interacting Quark-Gluon Plasma

My first indirect "encounter" with QGP happened, however, several years before this concept, and even QCD, existed. It goes back to 1971 when Ady Mann and I proposed[225] that Hagedorn's maximum temperature of hadronic matter was rather a critical temperature, characterizing a phase transition of that matter: at low temperatures hadronic matter is in a superfluid phase, and at high temperatures in another, not yet known phase. This yet unknown phase turned out to be QGP. Moreover it will be shown in the following that some of the other analogies discussed in this volume are in one way or another reflected in this new type of analogy: this refers among other things to phase transitions in general, equilibrium and preequilibrium, hydrodynamics, hot spots, quantum optics and Bose–Einstein correlations.

In 1980, soon after the phase transition from hadronic to quark matter (QGP) had become an accepted concept, Gerald Fowler and I[43] addressed the question what implications such a phase transition had for the Hagedorn exponential mass spectrum. By comparing at high densities the free energies of a hadronic phase with an arbitrary mass spectrum with that of an ideal gas,[226] we reached the conclusion that the hadronic mass spectrum could not grow with mass faster than a power and therefore an exponential mass spectrum was incompatible with the existence of a quark phase. This no-go theorem was of course not unexpected; it just confirmed Hagedorn's assertion that an exponential mass spectrum implied a maximum temperature beyond which no state whatsoever was possible. Actually, this main result of the Hagedorn bootstrap approach was the most puzzling aspect of his theory, since on the one hand a maximum temperature was an unknown concept (this was also the reason why many people, including myself, could not accept the Hagedorn conclusion) and on the other because the experimentally observed hadron mass spectrum was consistent with an exponential form, albeit the unknown widths of the higher resonances qualified this observation. However, subsequent developments of theoretical and experimental nature to be described below put in a new light our no-go theorem.

While our superfluidity conjecture had been formulated years before the advent of QCD and a fortiori of QGP, it turned out to have consequences for these subsequent developments. The latest of these is related to the most startling experimental observation made at RHIC, namely the observation that the quark-gluon plasma at present energies — albeit, apparently, a new state of matter, different from hadronic matter — is still strongly interacting, in contradiction with the expectation of asymptotic freedom. This surprising property of QGP had actually been predicted by us in a paper[227] presented at the Quark matter meeting of 1983.

To explain how this prediction came about I should remind (cf. chapter 13) that in 1976 Shalom Eliezer and myself reformulated[59] the superfluidity conjecture with the help of the sigma model, in an effective field theoretical approach, incorporating quarks. Up to a point this superfluidity model has strong similarities with the picture, which emerged quite soon after the conjecture of the existence of a deconfined quark matter phase became a subject of active theoretical and experimental search. Indeed, lattice QCD calculations suggested[228] the existence of three distinct phases: at low temperatures and densities we have hadronic matter, while at high temperatures/densities we have the deconfinement phase, where there exist only massless quarks and gluons. Between these two phases there exists a third phase in which the quarks are deconfined, but still have mass, and therefore the chiral symmetry is yet broken. Moreover, in this intermediate, deconfined phase there still exist pions, which are however massless and therefore have a phonon spectrum. That is, presumably, the supefluid phase described above (for further details cf.[222]). This suggests that what was termed in the seventies "superfluid hadronic matter" also contained traces of a deconfined phase.

Another message of the above scheme was that the masses of the quarks at high temperatures/densities vanish. This last point lead to the idea[229] that confinement, which corresponds to the opposite — low density — limit, can be described phenomenologically by making the mass of the quarks at low densities infinitely heavy. This idea received renewed attention due to the recent experimental results mentioned above (cf. the Phenix and Phobos RHIC White Papers[230]) indicating that in the quark gluon plasma phase possibly seen at SPS and RHIC energies deconfinement is yet far from being achieved. Another way of expressing this fact is to say that the quark-gluon plasma observed is (still?) a strongly interacting system (sQGP).[231]

Sometime around 1972 I discussed with Vicki Weisskopf (if I remember correctly the occasion was a colloquium Vicki gave at Indiana University) the idea of superfluidity of hadronic matter, which Ady Mann and I had been working on. I was happy to realize that he did not consider it too exotic and a bit later I was even happier to see that he (and his collaborators) had similar ideas: In a certain sense the MIT bag model[232] is a concrete realization of a hadronic superfluid system, in which the condensate is made of gluons and/or quark-antiquark pairs, which determine the vacuum energy of the bag. In the late 1970s, when lattice QCD was still in an incipient stage quark matter used to be described mainly by the MIT bag model and even now one finds in the literature applications in which the equation of state of hadronic matter is based on the bag model (cf. e.g. Ref. 233). In this model quarks interact with a constant effective coupling constant within a confining "bag" at constant pressure.

The MIT bag model, in which confinement appeared to be independent of the density of the system, was applied both to spectroscopy and to phase transitions. At least in the last case this model seemed to us unsatisfactory, because it ignored the bag-bag interaction. In phase transitions to the quark matter state, when nucleon bags are squeezed, this interaction appeared to be essential.

Anticipating Strongly Interacting Quark Matter; the Archimede Effect as a Model for Confinement of Quarks and Gluons

In Ref. 229, inspired by our previous work on phase transitions within the Goldstone mechanism, in which the mass of "free" quarks is infinitely heavy while at small distances (high densitites/temperatures) it is small or zero, we proposed a simple alternative model in which confinement was an explicitly density dependent phenomenon: The mass of the quark turned out to be inversely proportional to the quark density: $m \approx 1/n$. In a certain sense our model represented the analog of the Archimedes or buoyancy effect in which the effective mass of an object depends on the density of the medium in which it is immersed.

When applied to phase transitions from hadronic to quark matter, we found[234] that in this model asymptotic freedom was expected to appear at densities much higher than predicted by the bag model or other models on the market. This led us to entitle our contribution to the Quark Matter Conference 83 "How free is the Quark-Gluon Plasma?" It is rewarding to find that the newest data from the relativistic heavy ion accelerator RHIC at Brookhaven appear to confirm this expectation, which I had almost forgotten until I was reminded about it at the beginning of 2004 by Bill Zajc, the spokesman of the Phenix collaboration at RHIC. The corollary of the question asked above is: where, actually, begins asymptotic freedom?

The simplicity of the relation between the mass of the quark and the density has determined many people to use it in other applications, in particular in the study of strange matter, i.e. matter composed of strange quarks only. Given the new findings about strongly interacting quark-gluon plasma at RHIC one could foresee that this use will be intensified in future phenomenological studies of quark matter. The equation relating the quark mass to density could serve as a *theoretical laboratory* for the study of the effect of confinement. The usefulness of such theoretical laboratories was demonstrated lately by the Veneziano formula, initially proposed to describe resonance duality, but which unexpectedly lead to an independent derivation of the Hagedorn exponential mass spectrum and ultimately, and even more surprisingly, to string theory. The relation $m \approx 1/n$ also describes a dual property, quark duality, relating asymptotically free quarks to confined ones. What new surprises this could lead to is written (perhaps even literally) in the stars.

Our phenomenological approach to confinement proposed in Ref. 229 was generalized in Refs. 234 by including an effective one-gluon exchange and finite-temperature effects. Eventually an equation of state for the QGP was derived, which took into account the effects of confinement as defined above and which was used to study the phase transition between that state and the state of nuclear matter, described by an effective finite-temperature mean field equation of state (EOS). The results of this investigation showed that the effects of confinement persist *up to much higher* densities/temperatures than those derived without considering this effect: The velocity of sound u, e.g. differed appreciably from its asymptotic value $1/\sqrt{3}$ up to quite high temperatures (at $T = 700\,\text{MeV}$ one still found $u = 1/\sqrt{3.5}$). This value, which is even now in agreement with lattice results, shows that the QCD-lattice EOS contains what became later known as the strongly interacting QGP. This EOS was subsequently used in most hydrodynamical calculations of the Marburg group described below. In this way we could explain the heavy ion results obtained at the CERN SPS accelerator, which other approaches, that did not use such an EOS, could not.

28

Evidence for Quark-Gluon Plasma from Particle Physics

The history of the search for QGP is full of other surprises.[222] Among other things, contrary to what one might be tempted to think reading the current literature, the first evidence for QCD matter *in the laboratory* was reported not from heavy ion reactions, but from particle reactions, and that already in 1979, only four years after its existence was predicted. Independent evidence for QGP was subsequently found, again in particle reactions, in 1986. The evidence for QGP in particle physics is based on the success of the Landau hydrodynamical model in explaining various experimental data with an equation of state reflecting the transition from a quark-gluon plasma to a hadronic system. Besides that, the very fact that hydrodynamics is applicable presupposes a large number of degrees of freedom in the early stages of the evolution of the system. This again can be explained by assuming that in the initial phase the system consists of quarks and gluons. Last but not least, hadron production processes in hadron and lepton induced reactions are characterized by a universal production process, which suggests a common intermediate phase.

The Mystery of the Large Number of Degrees of Freedom

The phenomenological success of the Landau hydrodynamical model[111] of multiparticle production has constituted for decades a challenge for high-energy physics, because hydrodynamics is a classical theory and it assumes local equilibrium. Both these assumptions imply a large number of degrees of freedom, and in 1953, when Landau formulated his model, the only known particles were pions and nucleons, and the mean multiplicities in cosmic rays were, for present standards, quite low. Therefore the Landau model (and other statistical models of strong interactions) were considered up to the mid seventies as exotic approaches, outside mainstream physics. The bigger the merit of a few physicists like Carruthers, Feinberg, Minh, Shuryak and others who kept the interest in the Landau model alive.[235] In the West Carruthers' paper[45] has been particularly instrumental in this situation, because it pointed out that most of the multiparticle production data obtained at the CERN intersecting storage ring accelerator ISR could be explained by this model at least as well as by other, more fashionable approaches, like multiperipheral or parton models. Another important contribution of this paper was the distinction between "prematter" and hadronic matter. Carruthers introduces the notion of "prematter," as the "medium of highly compressed and energetic hadronic matter" to be distinguished from "matter, the stuff of which S-matrix theory is composed (the asymptotic region)."

The issues of local equilibrium and number of degrees of freedom in particle physics, which had constituted for decades the main conceptual difficulties of the hydrodynamical model got in the view of some researchers, including myself, a simple and convincing solution with the discovery of quantum chromodynamics and the possible existence of the new state of matter — the quark gluon plasma. At present energies in a nucleon-nucleon collision, instead of 4 degrees of freedom (proton, neutron, spin up and down) one can expect 12 N_f degrees, where N_f (the number of quark flavors) ranges from 2 for u and d quarks to 3 for u, d and s quarks. 12 comes from 2 (spin) × 3 (color) × 2 (quarks and antiquarks); in addition the gluons have 2 spin degrees of freedom and 8 color degrees, so that we have some 40 to 52 internal degrees of freedom (at still higher energies the inclusion of charm, bottom and top flavors raises this number to 88). Another circumstance, that facilitated the acceptance of hydrodynamical methods, which belong to classical physics, was the finding that QCD admitted classical solutions. Last but not least it was realized that the only chance of proving in the laboratory the existence of QGP as a *state of matter* was to use hydrodynamics in the interpretation of data. This follows from the trivial but sometimes forgotten fact that a state of matter is defined by an equation of state and there is no other way to get information about the EOS than by using hydrodynamics.

While with the advent of dedicated accelerators and detectors the search and study of QGP in heavy ion reactions has become mainstream physics, the situation is not so clear in particle physics. However the fact that even in nuclear physics the "public opinion" had to undergo a process of learning and understanding, which has not finished yet and the progress of which is also reflected in the series of special meetings dedicated to local equilibrium in strong interaction physics[125] and, of course, in the regular Quark Matter meetings, makes one believe that the evidence for QGP from particle physics will get in future the attention it deserves, the more so that this evidence preceded that obtained with heavy ions.

Some History

In his 1953 paper[111] Landau postulated that the EOS is that of an ideal gas. With this postulate he could explain the most salient experimental observations like the energy dependence of the mean multiplicity of secondaries produced, the single inclusive rapidity distributions, which were in a first approximation of Gaussian form, resembling in the central region a plateau, and the single inclusive transverse momentum distributions. This distribution is given by the characteristic exponential p_T distribution corresponding to thermal equilibrium at an effective freeze-out temperature $T_f = m_\pi$. It is known that this formula describes well the data up to transverse momenta of ~ 1 GeV. Moreover, there is no convincing and equally simple, alternative explanation for this experimental observation, which in some sense is direct evidence for thermal equilibrium in strong interactions. All this suggests that Landau's educated guess about the EOS was correct. Actually it was the *simplest* ansatz one could make if one considered the high-energy limit, where all masses are negligible, and it is a fortunate accident that the EOS of QGP coincides with that of an ideal gas. The correctness of Landau's conjecture is even more amazing if one takes into account that twenty years later it was found that the EOS of *hadronic* matter is much softer than that assumed by Landau (the velocity of sound is of the order of $1/\sqrt{6}$–$1/\sqrt{7}$, instead of the canonical value $1/\sqrt{3}$ postulated by Landau). And even more to the point, we know now that the velocity of sound is not constant, but changes with the energy density, because of the phase transition from QGP, at high temperatures, to hadronic matter at freeze-out.

In the pre-QGP era this state of affairs was, of course, not known and because of that, and also because of simplicity reasons, in the applications of the hydrodynamical model always a constant velocity of sound was assumed, in most cases $1/\sqrt{3}$.

An explanation why this constant value for u worked so well in p-p reactions was given in Ref. 236. In this reference the (one-dimensional) equations of hydrodynamics were solved exactly with a more general EOS, corresponding to an energy density dependent velocity of sound, for two different center of mass energies: $E_{cm} = 63$GeV and $E_{cm} = 540$ GeV. Three different variants for the EOS were considered, which all satisfied the following boundary conditions imposed by the assumption of the transition from QGP to hadronic matter: for small energy densities the velocity of sound was $1/\sqrt{7}$ and for large energy densities $1/\sqrt{3}$. The results for the rapidity and low transverse momentum distributions obtained with these variants were compared among themselves, as well as with those derived for a constant speed of sound $1/\sqrt{3}$. It turned out that the four different equations of state lead to quite similar results, all in rather good agreement with data. The above results strongly suggest that in the hydrodynamical expansion with a variable velocity of sound only the initial value of u matters and this explains the success of Landau's educated guess.[237]

Independent of this background of historical interest, we have to consider the results presented above at their face value: as evidence for QGP.

Large Transverse Momenta and Another Analogy: Leakage

Further support for this conclusion comes from large transverse momenta measured at the ISR. While the exponential spectrum describes extremely well the transverse momentum distributions up to $p_T \approx 1\,\mathrm{GeV}$, this is not the case at larger values of p_T, where deviations from this form were observed. Interestingly enough, these deviations, which correspond to the high temperature domain, not only do not contradict the Landau model, but can also be used to obtain information about the EOS. Actually they were interpreted as "evidence for a change with temperature of the velocity of sound in hadronic matter and of a phase transition from a strongly interacting hadron phase to a weakly interacting QCD phase" and published under this title in Ref. 238. As far as I can gather this paper constitutes probably the first phenomenological interpretation of experimental data based on QGP.

The experimental large p_T data show a striking deviation from exponentiality beyond 1 GeV/c; the logarithmic slopes significantly increase with energy. However at p_T larger than 5 GeV/c one observes a resumption of the exponential behavior with an energy-independent slope.

The interpretation of this observation proposed in Ref. 238 starts from the Pomeranchuk freeze-out ansatz mentioned above, which explains why the bulk of the produced particles have limited transverse momenta ($\langle p_T \rangle \approx 0.3\,\mathrm{GeV/c}$), characterized by an exponential thermal spectrum, corresponding to a Boltzmann distribution. However, because of the statistical nature of the process, emission at $T > T_f \sim m_\pi$ is not absolutely forbidden and this must lead to *leakage* of particles from the excited system, before the expansion has ended and equilibrium has been reached. This preequilibrium emission is known to take place also in medium energy nuclear physics. To describe mathematically this effect we used the Khalatnikov equation[239] for the relativistic hydrodynamical potential of the Landau model. This equation describes the time evolution of the temperature T of the fireball from its initial value T_{in} to its freeze-out value T_f. Taking the velocity of sound u and the normalization as free parameters and applying this formalism to CERN-ISR data one finds that from $p_T = 0.1$ to $\sim 5\,\text{GeV/e}$ the data are well described by the model with a value of the speed of sound u in the range $1/\sqrt{6.4}$–$1/\sqrt{6.8}$, which is compatible with values obtained for u from the hydrodynamical model, when analyzing rapidity distributions in p-p and p-nucleus collisions, and is also in agreement with theoretical predictions. The "large" p_T region (1–5 GeV/c) appears as a smooth and natural continuation of the "low-p_T" region, and only in the "very large" p_T regime does something "new" happen: In the region (5–15 GeV/c) the theoretical distribution becomes again a pure exponential, but the data deviate strongly from this asymptotic form as long as the speed of sound remains unchanged ($\sim 1/\sqrt{6.8}$). However, they are well fitted by an exponential with a higher initial temperature T_{in} ($\sim 5m_\pi$), corresponding to a speed of sound of the order of $1/\sqrt{3.5}$. This fact was interpreted in Ref. 238 as evidence for the fact that the sound velocity u is a (step?) function of temperature and that the high temperature region corresponds to a "different physical situation, probably a new phase."

Support for the above interpretation comes from the facts that: (a) The hydrodynamical model with only one free parameter, viz. u gives a consistent description of the p_T spectra over the whole energy range and in the entire p_T range, for twelve orders of magnitude in cross section. (b) This free parameter is already fixed to within a few percent of the fitted value by independent experimental facts like the rapidity distribution, which reflects the freeze out stage and thus corresponds to a value of u in the range $1/\sqrt{6}$–$1/\sqrt{7}$. The energy dependence of the multiplicity, which is determined by the entropy of the initial stage of the system, is consistent with a value of $u \approx \sqrt{1/4}$. c) The two values of u can be understood theoretically. The value $u \approx \sqrt{1/6}$ was derived[240] for a resonance gas corresponding to hadronic matter and the value $u = 1/\sqrt{3.5}$ follows from lattice QCD calculations, suggesting that the large p_T data from p-p reactions are evidence for a phase transition to quark-gluon plasma.

In Ref. 236 predictions were also made for the slope of transverse momentum distributions at higher energies, which could be checked at the CERN large hadron collider LHC which will start running in 2008 (cf. Fig. 3) and compared with those derived from a parton model.

The above picture got independent theoretical support from the calculation of the speed of sound quoted above[234] and which reflected the remnant confinement effects.

Multiplicity Distributions

Another type of evidence for quark-gluon plasma in particle reactionsbased on the equation of state and the Landau *model* comes from multiplicity distributions. This aspect of QGP search makes contact with our considerations on quantum optics applied to strong interactions (cf. chapter 21). Multiplicity distributions describe the probability to produce a certain number n of particles in a given reaction. The dependence on the incident energy of the *mean* multiplicity of produced secondaries has constituted one of the first indications that the equation of state of hadronic matter in its initial stage corresponds to an ideal gas. Once reliable information about the *fluctuations of multiplicity*, reflected in the multiplicity distributions, became available it was natural to ask whether these distributions supported this conclusion about the equation of state.

The answer is positive and can be resumed in the following way[241]: the observed dependence of multiplicity distributions on energy and shifts of rapidity bins can be explained by assuming the existence of two sources. One source is concentrated at small rapidities, has properties of a thermally equilibrated system, as could be expected from a quark-gluon plasma and gives rise to a distribution of negative binomial form; the other one is contributing to the whole rapidity region, displays characteristics of bremsstrahlung emission and has thus the form of a Poisson distribution. With increasing energy the weight of the "thermal" source increases.

Global Equilibrium: The Universality of the Hadronization Process

If "prematter" as defined by Carruthers really exists then it should manifest itself not only in reactions induced by hadrons, but in other hadron producing reactions as well. This has been rather convincingly proven in e^+-e^- reactions. For this purpose the energy available for hadron production in p-p reactions has been compared with that in e^+-e^- reactions: It appears that one and the same hadron production mechanism is at work. This strongly suggests that although the initial state is different in the two cases, there exists a universal intermediate state,[242] common to both systems. The universality manifests itself, among other things, in the energy dependence of the mean multiplicity, in the inclusive distributions and last but not least in the ratios of particle species of secondaries. Not only are these ratios, considered separately for hadron-hadron and e^+-e^- reactions, in accordance with what one would expect from systems in thermal and partial chemical equilibrium at a given freeze-out temperature, but this freeze-out temperature is a universal constant, independent of the initial energy and the initial type of reaction. It is not difficult to guess what this universal intermediate state is made of: quark-gluon plasma.

Atomic Mass Dependence of Multiplicity

The amazing success of the Landau hydrodynamical model in particle reactions, as briefly sketched above, is matched by various successes in nucleus-nucleus (A-A) reactions, which will be discussed below. While in nucleus-nucleus reactions the issue of the number of degrees of freedom is not directly at stake, there exists at least one prediction for A-A collisions, which lends independent support for the idea that the number of degrees of freedom is big enough in *particle reactions* as well, to justify the application of hydrodynamics. This is so because this prediction follows straightforwardly from the hydrodynamical theory of p-p collisions. Since this prediction is based on the QGP EOS it constitutes independent evidence for QGP in particle reactions: We refer to the calculation of total multiplicity N for central A-A reactions performed by Landau already in his first paper[111] and which was found, already in the 1950's, to be in approximate agreement with experimental cosmic rays data. This agreement was confirmed by subsequent accelerator particle data and by most recent heavy ion data.

At this point a short comment on the relevance of particle physics as a testing ground for QGP may be appropriate. At a first look it seems that particle reactions present two major disadvantages in this respect:

(a) the number of degrees of freedom;
(b) the possibility to see deconfinement over distances exceeding the nucleon size.

These were the most important reasons why heavy ion reactions have become the main tool in the search for QGP. Unfortunately, despite special signatures designed to prove *extended* deconfinement, this goal has not been convincingly achieved either at the CERN SPS or at RHIC, and it is questionable whether one might be ever able to demonstrate extended deconfinement in the laboratory by the methods envisaged so far. Surprisingly enough, particle physics itself might help surpass this difficulty. Indeed, if QGP has been seen in p-p reactions and if certain variables (like multiplicity e.g.) in A-A reactions scale with respect to p-p reactions then this strongly suggests that these variables reflect QGP behavior over nuclear sizes. It is an even bigger surprise and a strange irony of history that the apparent disadvantage (a) turns out to be, against all expectations, rather an advantage in the search for QGP: While in heavy ion reactions the number of degrees of freedom is from the beginning not at stake — the factor *A* obliges — in p-p collisions it has become an important piece of evidence for the existence of QGP. These considerations confirm the importance of particle reactions in the search for quark matter.

29

Evidence for Quark-Gluon Plasma from Heavy Ion Reactions

Traces of QGP in Low Energy Heavy Ion Reaction?

The search for quark-gluon plasma in the laboratory is usually considered a topic of high energy physics. However *high* energy is a time dependent concept reflecting the status of accelerator technology. Actually it is conceivable that manifestations of quark-gluon plasma have been seen at energies much lower than those of present heavy ion facilities. Simple estimates of energy densities suggest that constituent and current quark deconfinement could take place even at Bevalac (1.7 A-GeV) and Dubna (4.5 A-GeV) energies. In the following I shall address one particular effect of this kind, because it is also related to the ongoing investigations of elliptic flow at RHIC: I refer to the shear viscosity of matter. In most applications of hydrodynamics to high-energy reactions viscositywas neglected, i.e. the fluid was considered ideal. The success of hydrodynamics in the pre-RHIC era justifies this assumption. For systems with phase transitions finite size and surface effects have to be considered. Considering non-central heavy ion collisions in which the "participants" may be heated up and undergo a transition to the QGP, while the "spectators" remain in the hadronic phase, it was suggested by Halzen and Liu[243] that the shear viscosity between the two phases is small. This is so because the exchange of particles between these phases, one of which is in a locally colored phase, is (almost) negligible and limited to the surface separating the two phases. Solving the equations of hydrodynamics in the transverse plane with viscosity (corresponding to the case when no quark matter is formed) and without viscosity, (corresponding to the case when there is a phase transition) the authors[243] found a *separation in rapidity* of the two phases: in the projectile frame the rapidity distribution of the particles produced from that part of the system, which underwent the phase transition, is shifted towards smaller rapidity values as compared with the distribution of those originating from the part unaffected by the transition.

As far as I can gather this effect has not yet been looked for. An obvious difficulty in searching for it is that the separation in rapidity is a phenomenon, which could be attributed to other causes as well: Rapidity gaps appear in diffraction scattering and jet production, to mention just two more common effects. There exists, however, an alternative signal, also based on the reduced value of the shear viscosity between QGP and hadronic matter, and it is conceivable that this signal proposed in Ref. 244 has been seen.

While Halzen and Liu assume a high degree of transparency, probably not yet reached even at RHIC, our considerations[244] apply to much lower energies, where stopping dominates and where a clear distinction between participants and spectators can be made. This means that in each event there are (a) one or two highly excited fireballs, which contain a lot of baryons and which contribute to the central rapidity region, and (b) two low-temperature fireballs populating the non-central (fragmentation) regions. Due to fluctuations the phase transition does not take place in each event. In those events where it does not occur, viscosity contributes to the heating of the system and the "low" temperature of the fragmentation region will be higher than in those events where quark matter is formed.

It is amusing to mention that such two temperature events have indeed been seen in two independent emulsion experiments[245,246] with 1.7 A-GeV Fe beams and subsequently also[247] with 4.5 A-GeV ^{12}C beams. In these experiments the transverse momentum spectrum of α particles in the projectile fragmentation region was measured. In Ref. 246 it was found that this spectrum is characterized by two effective temperature of 10 and 40 MeV respectively. This was confirmed in Ref. 246 with the important specification that these different temperatures belonged to different events, which was interpreted as evidence for different reaction mechanisms.

An Excursion Into Cosmology

The preceding considerations refer to the laboratory evidence for QGP. Before continuing on this theme I would like to make a detour into cosmology, because our work in Marburg on high-energy heavy ion reactions has actually started by this detour. Standard cosmology tells us quite convincingly that soon after the Big Bang the universe consisted of hot quark-gluon plasma, which, by cooling, became hadronic matter. As a matter of fact, this has been an important argument for convincing the scientific community that the search for QGP in the laboratory is a worthwhile enterprise. This cosmological evidence as well as the circumstance that the search for QGP concentrated in the 1980s almost exclusively on heavy ion reactions determined me to shift our hydrodynamical studies to high-energy nuclear-nuclear collisions, the more so that this topic became in Germany "mainstream" physics and it was easier to get, from the funding agencies, support for graduate students.

Easier, however, does not mean easy. This was not so much due to the scarcity of funds, than to another circumstance: This field was interdisciplinary, at the interface between particle and nuclear physics, and there were very few referees — and this is an understatement — competent in both fields. The number of competent phenomenologists was even smaller.

In a "warming-up" exercise,[248] which has found recently some resonance in the literature, we used our previous experience gained in the studies of the influence of the equation of state on the hydrodynamical expansion. That the search for quark-gluon plasma via hydrodynamical studies is relevant for the understanding of the evolution of the universe after the Big Bang is not only due to the fact that at a certain stage the universe consisted of a QGP, but also, and most importantly, to the fact that the expansion of the universe is described by hydrodynamics, as well. It was then quite natural to ask how the details of the equation of state and of the phase transition influenced this expansion.

In 1986–1987 Udo Ornik, who had just joined the Marburg group and worked on his PhD thesis, and I tried to answer that question. This study was of interest for us also from a technical point of view, since in this case one-dimensional hydrodynamics is not an approximation, but rigorously valid: There is no initial privileged direction and the expansion occurs radially. Another difference between this case and that encountered in particle reactions is that for cosmology one has to consider also the Einstein gravitational equation. For the sake of generality and inspired by our previous studies we assumed that the velocity of sound depended on energy density; furthermore we considered different variants of the equation of state, including phase transition of various order. Rather than assuming two different equations of state for the different phases, we used one single equation of state, which contained the two phases. From a theoretical point of view this procedure presented progress and it constituted the departure point for the subsequent systematic investigation of heavy-ion reactions. However, for these reactions we could not limit ourselves to one-dimensional solutions, but had to solve the system of equations in 3+1 dimensions.[249]

Solving Exactly the Equations of Hydrodynamics

As emphasized before the equations of relativistic hydrodynamics being non-linear, their solution is a non-trivial mathematical problem. Exact analytical solutions of the equations of hydrodynamics for arbitrary velocities of sound u exist only for the [1+1] dimensions case, and that only for constant u. (For p-p reactions it is the Khalatnikov solution mentioned above). For the more complex case of p-A reactions an analytical solution was found in Ref. 250. That is why, starting with Landau himself, analytic approximations were used or, alternatively, exact, but numerical solutions.

The mathematical complexity of the problem we faced was enhanced due to the *double non-linearity* of the equations: not only were the equations of hydrodynamics non-linear, but the equation of state was non-linear as well, given the energy density dependence of the velocity of sound. Another complication in the application of hydrodynamics is represented by our limited knowledge of initial conditions. This explains why for heavy ion physics some authors preferred to *interpret* physical observables in terms of hydrodynamical and thermodynamical concepts like flow and temperature, rather than to *derive* them from the equations of hydrodynamics. This approach is obviously unsatisfactory, since it throws the baby out with the bath: with such an approach one cannot obtain information about the initial state and the different phases of the system (i.e. the equation of state), which we are really after.

With the advent of powerful computers exact numerical solutions in [3+1] dimensions (for central collisions cylindrical symmetry reduces the [3+1] problem to a [2+1] problem) have become more and more used and this effort has paid. Among other things, in the period 1989–1996 the Marburg group obtained and applied to the CERN heavy ion high-energy data[251] an exact numerical solution (Hylander) of the equations of hydrodynamics for [2+1] dimensions, *without assuming boost invariance*. The EOS used was taken from lattice QCD (LQCD) results: it contained a phase transition between a hadron gas and a QGP. It is important to note that according to LQCD this QGP is, at the temperatures considered, not yet an ideal gas, the corresponding velocity of sound at $T = 300\,\text{MeV}$ still being around $1/\sqrt{4}$ instead of the asymptotic $1/\sqrt{3}$. This fact confirms the strongly interacting character QGP at the energies available so far, both at SPS and RHIC.

The evidence for QGP from heavy ion reactions, which the Marburg group found, is based on the simultaneous comparison with SPS data of the theoretical results for single inclusive rapidity and transverse momentum distributions, as well as of the Bose-Einstein correlations. These theoretical results were obtained from the exact solutions of hydrodynamics with the LQCD equation of state, which contained the phase transition from hadronic to quark matter. *This comparison showed that hydrodynamics with an EOS that included the phase transition could explain all the data, while hydrodynamics, with the same EOS but which did not include such a phase transition, could not.*

Many of the experimental features explained by this hydrodynamical approach were actually predictions, which were later confirmed in experiment. Not only is this one of the earliest pieces of evidence for QGP in heavy ion reactions at SPS energies recorded in the literature, but it is from a certain point of view superior to other indications[252] for QGP considered up to recently as "standard," because of its EOS relevance.

The fact that Landau hydrodynamics "rules the waves" not only in p-p reactions, but also in A-A reactions, was for many a surprise, because of the erroneous, but dominant belief in boost invariance, which turned out to be in contradiction with data. But this was not the only surprise: The non-linearity of hydrodynamics compounded by the non-linearity of the EOS produced another surprise, which contradicted naïve intuition and "predictions" based on this intuition. Indeed, besides taking into account partial stopping, the initial conditions assumed in Ref. 251 were such that on top of the longitudinal expansion there was, from the very beginning, a transverse expansion as well. The comparison between theory and experiment shows that, contrary to what was assumed until then, the transverse expansion plays from the very beginning an important role. This is seen in particular in Bose-Einstein correlations, where in opposition to what one might expect intuitively, the transverse flow does not increase the transverse radius; if anything, it rather contributes to its decrease.[253] This effect is a consequence of the fact that with increasing time the longitudinal coordinate of the points of freeze-out increases, while their radial coordinate decreases.

Outlook

We arrived at the prediction that QGP may continue to be a strongly interacting system at temperatures much bigger than the critical QCD temperature starting from the superfluid properties of hadronic-partonic matter as seen in *particle* reactions. Twenty years later sQGP was experimentally discovered in *heavy-ion physics* at RHIC. This lends independent support to the superfluidity hypothesis and the profound link between high-energy particle and nuclear physics.

One might wonder whether superfluidity may not provide also an explanation for the apparent ideal character of the hadron-parton fluid seen in strong interactions. (The experimental evidence for this ideal character is based, among other things, on the success of ideal fluid hydrodynamics in its application to multiparticle production phenomena in p-p, p-A and A-A reactions, the observation of hot spots, which can be produced only if heat conductivity in hadronic matter is small, and last but not least on the strength of elliptic flow seen at RHIC, another of the real surprises of RHIC). Since a superfluid has by definition zero shear viscosity one might expect that a system, which contains a mixture of a normal fluid and a superfluid is more ideal than the purely normal one. Yet, whether this expectation is also fulfilled for a rotating system, such as that encountered in elliptic flow, is unclear. In a rotating bucket containing *superfluid helium II* for example, it is not fulfilled: vortices arise, which make the entire liquid move with the bucket. However, another superfluid system made of cold, strongly coupled atoms apparently behaves like a perfect fluid.[254]

It might be amusing to note that the strongly interacting character of QGP discovered at RHIC has in a certain sense also produced deceptions. Indeed, while in 2000 spokespersons from the experiments on CERN's Heavy Ion program presented "compelling evidence for the existence of a new state of matter in which quarks, instead of being bound up into more complex particles such as protons and neutrons, are liberated to roam freely...,"[255] in 2005 the four main experimental groups of the Brookhaven accelerator RHIC were much more prudent in their White Papers,[256] where they presented their conclusions with respect to the search for quark-gluon plasma: they agreed with the CERN statement only in the sense that a state of dense matter had been observed in heavy ion reactions, which was difficult to explain in terms of hadronic degrees of freedom.

In my view the explanation for this change of tone is the following: if the matter we see at RHIC is strongly interacting, one might ask what is the difference between this matter and conventional nuclear matter? Where is the evidence for a *new* state of matter? Up to now the main qualitative signature of QGP was thought to be its weak interaction and this signature has suddenly been lost. Equally embarrassing might be the repercussions of this state of affairs for the proof of extended deconfinement, another key characteristic of QGP: Given the strongly interacting character of the matter found at RHIC some people are now asking whether the question of extended deconfinement makes even sense for such a system. On the other hand the fact that the strongly interacting character of QGP had been predicted before its experimental observation could serve as an argument in favor of the interpretation of the dense matter found at RHIC as genuine QGP.

We have seen that in the search for quark gluon plasma, like in any research field, there have been real and less real surprises. For many a surprise may have been that the evidence for QGP is probably much older than usually thought. Another surprise is the breakdown of intuition. To quote just two examples, which illustrate this last point: transverse expansion does not increase the transverse radius, and asymptotic freedom and the fact that the velocity of sound at present energy densities has reached almost the ideal gas limit does not mean that the QGP is a weakly interacting system. Both these examples are a reminder of the limitations of analogies.

30

A Literary Intermezzo: The Mini-Atom Project

Besides the scientific projects pursued after moving to Paris, around 2000 I also started to write a detective story with scientific background, the first version of which was finished around 2002. Why did I do that?

After more than 50 years of physics I felt the need for a change, the need to relax a bit. Furthermore, everything I had done or written so far interested (if at all) quite a narrow audience, the audience of physicists specialized in the fields I had been working. An attempt to go beyond that, an attempt to change the audience, would certainly constitute a change. I was then faced with two possibilities: 1) to explain what and why I had done in physics to non-specialists and even to the public at large. This possibility lead to the present book; it is in part the realization of the fact that most of the topics I had been working on in my career as a physicist were analogies; 2) to write a detective story with scientific background. Indeed, research is a kind of detective story; you try to solve a puzzle. Furthermore, I had a physics idea, which, however, I was unable to prove. The combination of these two elements led necessarily to a thriller with a slight science fiction flavor.

The idea has to do with the constants of nature. Physicists have long since asked whether the various constants of nature like speed of light c, the Planck constant h and the various coupling constants, including in particular the fine structure constant, which combines the electromagnetic coupling constant, i.e. the elementary electric charge e, with c and h, are really constant. Actually, in modern field theories — gauge field theories — the coupling constants are "running," i.e. depend on the energy. This has been experimentally confirmed both in electromagnetic interactions and strong interactions. The electromagnetic coupling in particular increases with energy. While at present energies this change is quite small, at very high energies it must become appreciable.

The electromagnetic coupling determines the size of atoms: the bigger, that is the stronger the coupling, the smaller the radii of atoms are. Furthermore, large energies are equivalent to high temperatures (this is the principle on which the investigation of quark matter in the lab is based). This suggests that at high temperatures the sizes of atoms will be smaller than those we observe at present. If such conditions could be created in the lab atoms could be transformed into mini-atoms, and normally sized objects into mini-objects.

The hero of the novel, Trevor, is a theoretical physicist who not only discovers this possibility, but also a physical mechanism through which it can be implemented. The mechanism is a chain reaction, which is entertained by converting mass into energy. The proof that this mechanism works involves very long and complex computer calculations, which are very demanding. In the meantime the American, Soviet, and French intelligence agencies, which have become interested in the project, because they think it could lead to military applications like mini-atomic bombs and mini-rockets, pursue him. A hacker in the service of the KGB penetrates into Trevor's computer files. This event, the possible destructive applications of his idea, of which he has become aware in the meantime, and the computational difficulties Trevor meets in solving his problem, contribute to create a love-hate relationship between him and the computer. The computer becomes for him the personification of good and evil and he starts to see in himself a Dr. Jekyll-Mr. Hyde. He dies eventually, under mysterious circumstances, by his computer.

Unlike the present book, in writing the novel I was faced with the terrible question what language to choose. I was in the definitively unlucky situation that I had a choice between English, the language in which all my publications were written and in which I would find a broader readership, and German, my native tongue and the language of the country in which I lived for the last 25 years. And for the writing of a novel this last circumstance is quite essential. Since I could not make up my mind, I alternated the two languages; some chapters were initially written in German and then translated by me into English and vice-versa. From the literary point of view this is certainly not a very satisfactory procedure and that is why I considered myself unlucky. On the other hand because of this "indeterminacy" I had from the outset the novel in two languages. The German version appeared in 2006[a] and had quite an unexpected impact. It got very favorable reviews in prestigious newspapers like *Neue Nürcher Zeitung* (31 October 2007) which compared Trevor with a character of a novel by Stefan Zweig and *Frankfurter Rundschau* (14 February 2007), which also published my photograph. I was also invited to the "solo" high-brow 50 minutes radio emission *Doppel-Kopf* of Hessischer Rundfunk 2, in which in the past participated leading writers, artists, and other high-society personalities like Hans-Dietrich Genscher, the former German Foreign Affairs minister, Rita Süssmuth, the former President of the German Bundestag and Minister for Social affairs. The English version is not yet published.

[a] *Das Miniatom-Projekt*, Literatur Wissenschaft.de (TransMIT) 2006.

31

Reflections

Breakdown of Analogies: Mysteries of Modern Physics

In this book I have been mostly concerned with the use of analogies for the prediction or explanation of new phenomena. We have seen that analogies between phenomena in different chapters of physics have been an essential and quite general method in the evolution of physics (the situation is quite similar in other branches of science). Any analogy, however, has its limitations. The elucidation of these limitations is as important as the formulation of the analogy itself, since it leads to new findings.

In the history of physics it sometimes happened that analogies lead to a dead end; they proofed to be inapplicable. These breakdowns of analogies, which often appeared as "mysteries" or even crises, in their turn often lead to the discovery of new phenomena and even of new chapters of physics. The discovery of quantum physics is one of the most characteristic examples in this category. It owes its existence to the failure of classical mechanics in atomic physics and in particular to the breakdown of the planetary model of Rutherford.

Classical Versus Quantum Physics

When judged by our day-by-day experience the world of elementary particles is full of such mysteries, which appear in the form of restrictions and which reflect the difference between the laws of classical physics and quantum physics. The Heisenberg indeterminacy relations, which prevent in quantum physics the simultaneous determination, with arbitrary precision, of a pair of certain variables, e.g. coordinate and momentum of a particle or energy and lifetime of a physical state (cf. chapter 5) illustrate quite vividly this point. Such a restriction does not exist in classical physics.

Another type of mysteries specific for quantum physics is the principle of identity of particles: While in our macroscopic world we always can distinguish between two objects, however much identical they might appear, this is not possible with quantum systems. Two drops of water or two grains of sand, when looked at with a magnifying glass, still present differences in size or weight, even if very tiny, and thus can be distinguished. Moreover, if we baptize at a given moment two particles that appear identical by the letters *a* and *b*, in classical physics we are always able to pursue this identification at later moments and tell what happened with particle *a* and what happened with particle *b*. In quantum mechanics this identification is not possible, the two particles loose at later moments their identity and are indistinguishable; we cannot tell which is (was) *a*, and which is (was) *b*. This is true for identical elementary particles like electrons or protons, but also for identical systems made out of these, like atoms or molecules.

The principle of identical particles is related to another mystery restriction — that of the number of particles in a given quantum state. While in classical physics the number of particles in a given state is arbitrary and does not depend on the nature of that particle, in quantum physics this is not the case. Not all species of identical elementary particles can occupy one and the same quantum state.

The question how many identical particles can be in a given quantum state depends on the total angular momentum (spin) of the particles. Identical particles or systems with integer spin, *bosons*, can be in the same quantum state, while particles with non-integer spin, *fermions*, cannot. The presence of one fermion of a given type in a given state, excludes the presence of another one of the same type in that state. This *exclusion principle,* discovered by Pauli, explains, among other things, the stability of atoms and atomic nuclei: the constituent fermions (electrons in atoms and nucleons in nuclei) are organized in shells, which assures the stability of these systems (cf. chapter 5). From this point of view bosons are more classical than fermions. Mathematically this is reflected in the fact that the classical limit of a Fermi field is not an ordinary c-number but an anticommuting q-number, satisfying a Grassman algebra. This fact is also reflected in the relationship between classical fields and coherent states. Coherent states are eigenstates of the annihilation operator and therefore classical fields are always coherent states: if one annihilates one particle out of a classical field, this field is recovered, since a classical field contains an indefinite number of particles.

The difference between bosons and fermions manifests also in the symmetry properties of the wave function of the system under permutations; for bosons, which satisfy Bose–Einstein statistics, the wave function is symmetric under permutation of two identical particles, while for fermions the wave function is antisymmetric; fermions satisfy Fermi-Dirac statistics. In quantum field theory this property is assured by the commutation of Bose field operators and anti-commutation of Fermi field operators.

While the elucidation of some of the mysteries mentioned above is still a task for the future "theory of everything," the investigation and use of their consequences has been the subject of modern physics commencing with the beginning of the 20th century. The fact that bosons of the same type can be "packed" together in one state makes possible the formation of Bose condensates, in which the number of identical particles in a given quantum state is indefinite. These condensates, despite being quantum systems, are of macroscopic size and have important practical applications: Superconductors made of these condensates serve, e.g. in modern accelerator technology. That this technology serves in its turn to study particle physics is another illustration of the link between macroscopic and microscopic physics.

Finally, to end this enumeration of unanswered questions, which anyway has no pretension to be complete: Why do the constants of nature have the particular values we observe in experiment and not other values, why e.g. is the velocity of light in vacuum 3×10^{10} cm/s, the Planck constant 2×10^{-23} MeV, the mass of the electron 0.5 MeV, the fine structure constant 1/137 etc? Although the standard model represents a huge progress in the quest for understanding the microcosm it has not provided an understanding of the mysteries mentioned above. Moreover it has brought up some fundamental questions of its own, which wait to be answered. One of these is related to the quantitative question posed above: Where from comes the fine-tuning between the parameters of the standard model, which assures its consistency?

Among the qualitative questions unanswered by the standard model one could mention: Why do strong and electromagnetic interactions conserve parity while weak interactions do not? Why are leptons (or baryons) fermions and not also bosons? Why do there exist three generations[a] of elementary particles?

Within the standard model we have been only able, so far, to *describe* these facts, to *organize* this empirical information. Nobody has been able to *answer the above questions*, some of which, as we have seen, have been put already before the advent of the standard model.

[a]Generations (or families) of particles are sets of quarks and leptons with identical quantum numbers and interactions, which differ by their masses.

Intuition and Analogies: Three Definitions of Intuition

From the facts quoted above follows that the laws of quantum physics are fundamentally different from those of classical physics and we have to adapt our way of thinking, in other words our intuition, accordingly. When I use here the word intuition I have in mind its meaning in natural sciences, which does not fully coincide with its meaning in philosophy. The last one is defined in the Encyclopedia Britannica, as "the power of obtaining knowledge that cannot be acquired either by inference or observation, by reason or experience." For scientists intuition has a somewhat different meaning. I discussed once with Salam this problem and he proposed a very simple and short definition of intuition: "Intuition is the background that an individual has."

I would like to suggest here another definition of intuition, which is consistent with Salam's, but makes it more specific: *Intuition is the understanding based on analogies.* The breakdown of our intuition in the domain of quantum physics, as manifested in the above mysteries, is an example of the limits up to which analogies are permissible and useful.

The above three definitions of intuition are not equivalent. Among other things, they vary with respect to the degree of subjectivity they allow for. The definition used in philosophy and that of Salam emphasize the subjective nature of intuition. According to these two definitions what appears to one individual more intuitive, to another might appear less. On the other hand the definition in terms of analogies gives intuition a rather objective character. This makes it more appropriate for sciences of nature and physics in particular. To illustrate this let us consider the mysteries of quantum physics discussed above.

The indeterminacy relations, as discussed by Heisenberg himself, can be explained up to a given point by arguing that to measure the position of a particle that has a well-defined momentum we have to look at it under a microscope. However this implies that we have to submit it to the action of photons and these will change the momentum of the particle. With this analogy the mystery of the Heisenberg indeterminacy relations is intuitively understandable. This is less the case for the principle of identical particles and even less for the mystery, which relates spin with statistics, as it is quite difficult to find an analogy in classical physics with the notion of non-integer spin. (That is in line with the observation that bosons are more classical than fermions). I think that (almost) everybody will agree with this hierarchy of intuitiveness.

Referees, Editors, and Publishers

Most of the physics content of this book was dedicated to the process of insight into, and prediction or explanation of new phenomena. But this is not yet the end of the process of research: To convince yourself and your peers that your findings are correct you have to make your results known. This is usually done through publications. This stage has its own difficulties and I will try to sketch below some of the experiences I gained about this stage in the fifty years of my scientific life. These experiences may be perhaps useful for those who are at the beginning of this road.

Original Papers

My experience with scientific publications is limited to physics, however I understand that the situation is similar in other scientific disciplines.

To get a paper accepted for publication in a scientific journal, it has to be refereed by at least one referee. (In the case of *Physical Review Letters* positive reports from at least two referees are necessary). The same applies for scientific books. The referees are appointed by the editors. One may ask who are these referees and most importantly, how does the editor decide which referee should be consulted for a given paper.

As a rule, referees are peers of the author who have already published in that particular field and the publications of which have had a certain resonance. (This distinguishes science from art where critics themselves are seldom artists). Lately this resonance is measured by the frequency of quotations, although that criterion can be misleading. More delicate is the question of the choice of the referee. For a theoretical paper on a subject in which there exists two or more schools of thought, that choice can be decisive. If the referee does belong to the same school as the author, then the paper has a fair chance of being accepted, provided that it does not show obvious errors or make claims that are too strong. If this is not the case, an author can have a hard time to get his paper published.[b]

I remember a discussion that I had on this subject in 1970 with Sam Goudsmit, at that time the editor of *Physical Review Letters.* I was visiting Brookhaven National Laboratory, the headquarters of *Physical Review Letters* and I was curious to meet also the (co)discoverer of the electron spin and the leader of the Alsos[258] mission. Moreover, I had also published in Goudsmit's journal and had been sometimes puzzled by the comments of the referees of my papers: it seemed to me that some of them were completely outside the subject. Goudsmit's reaction was quite candid and disarming: "in the future please submit to me with the paper also a list of ten people whom you believe knowledgable in that field and I assure you that one of them will be the referee." I consider this approach highly recommendable, because an author will accept much more easily comments and criticism from colleagues he believes to be experts, than from unknown referees, the competence of which might appear to him doubtful.

[b]Some editors reduce the criterion "same school" to geography and this can be quite misleading as well.

As a matter of fact, I myself have had the opportunity to experience in a more direct way the convincing power of the knowledge of the name of the referee, when after submitting a paper to *Physical Review Letters* I got directly from one of the referees, with whom I had corresponded on the topic, a letter with comments and suggestions of improvement. It goes without saying that I accepted and implemented these comments immediately and indeed got the paper published. That referee was Harry Lipkin and later on, being myself a referee, I practiced from time to time that procedure.

Among the physics journals, those of the American Physical Society play a particularly important role, because of their quality and prestige and also because the APS is a non-profit organization. That last point is the reason why there are some formal rules the editors of these journals have to obey in their relationship with authors, including the right of appeal and the right to know the comments of the referees. Most other journals apply now similar rules and all overall, I have the impression that at present it cannot happen that an author of a paper that is at least *apparently* scientifically correct and reasonably written, should not be able to get it published. In case that one journal rejects it he always has the possibility to submit it to another journal and in each field there are several journals. Some people may consider that unfortunate because it contributes to a proliferation of papers, many of which are unnecessary; that makes difficult to follow the developments in a given field. Another consequence of this multiplication phenomenon is the (mostly involuntary) repetition of results. Of course, the proliferation phenomenon is in part also a reflection of the slogan *publish or perish*.

The internet, where everybody is free to make his paper available to the public, has acerbated this phenomenon, but there at least the reader knows that the paper was not refereed.[c] An experienced

[c]This does not apply to *internet journals* which use a referee system similar to that of printed journals.

scientist will usually consider the results of an internet paper with that caveat in mind.

Review Papers

While there is an abundance of journals publishing original papers, with review papers the situation is rather different, one can hardly sustain that there are too many of these. The increasing need of review papers is a consequence of the multiplication of "original" papers, which makes it difficult to follow the developments in a given domain. Publishers encourage therefore the publication of review papers, by often paying a (nominal) fee and sometimes by inviting such papers. Review talks presented at meetings fill in, in part, this gap, provided they are published in the proceedings. However, besides the facts that review talks are usually not written with the same care as review papers and that they have to be much shorter, they are not easily accessible, because, as a consequence of the catastrophic financial situation of libraries all over the world, the proceedings are rarely available in department libraries. When the library funds are cut, the first thing that one renounces is the acquisition of books and proceedings and one keeps the subscription of journals. That is why review papers published in review journals are in principle more accessible. However the continuous deterioration of library funding have put also this statement under question mark, because in the last two decades libraries started to cut even the subscription of journals. That was one of the reasons why I decided to move to Paris in 1995. Although French universities and research institutes went through the same financial crisis as those in Germany, given the large concentration of physics institutes in Paris the situation was less dramatic: one still can find in the Paris region practically all important physics journals.

Fiction

There is a fundamental difference between the process of publishing of scientific work and that of literature: a scientific work is always refereed by a scientist, who is also an author of scientific papers. A literary work is refereed by a literary critic who, in general, is not a writer himself. Moreover the language a critic uses is often different from the language writers use and this explains why readers often do not understand what a critic wants to say, while they usually are able to form a definite opinion about a literary work they have read. One often has the impression that critics address themselves less to the reader or author than to their colleagues. Writing literature is a different profession than doing literary criticism. Reich-Ranicki, a famous contemporary German literary critic, satirizes this state of affairs by saying that writers have no idea at all what literature means.

The divorce between artists and critics is not restricted to literature, but applies to other artistic fields as well.

Appendix 1

Statistical Duality

As mentioned in Chapter 9 the Regge model predicts only the energy (s) dependence of the differential cross section of these reactions; the dependence on momentum transfer (t) (contained in the residue of the Regge pole) is left undetermined. Moreover, even this prediction is based on a relationship between spin and mass, which is an extrapolation involving complex angular momentum, analyticity and crossing symmetry. Some information about the t dependence of cross sections can be obtained from finite energy sum rules and duality: The amplitude for a given process is supposed to be "saturated" by resonances in the direct (s) channel, in the sense that the integral (in s) over the (imaginary part) of the resonance scattering amplitude is numerically equal ("dual") to the sum over all Regge amplitudes. The t dependence is contained in the (unknown) Regge residues of the left hand side of this equation.

The statistical mesonic-cloud model, which leads to the conjecture of superfluidity of hadronic matter, provides another relationship between the s and t channels. Indeed one can show[259] that, under certain plausible assumptions, by integrating the differential cross section as given by the mesonic-cloud model over all physical t values, one gets an average, s dependent, cross section, which has the smooth power dependence characteristic for Regge behavior. This is reminiscent of finite energy sum rules and duality and because cross sections rather than amplitudes were involved, I called this relationship *statistical duality*.[259] In this sense the s dependence of cross sections, a prediction, which had been considered until then characteristic for Regge theory, can also be considered as a consequence of the mesonic-cloud model. This result is consistent with the fact that, as mentioned in chapter 12, resonances can in some sense be viewed as a consequence of the superfluid properties of hadronic matter and which were also derived from the mesonic-cloud model. In this context one should mention that within the above approach elastic scattering (Pomeron exchange) is due to phonon exchange. Finally another consequence of this model is that at large t the differential cross section cannot decrease with t faster than a power and therefore the observed exponential behavior of the differential cross section at small t, which is a consequence of the Bose-Einstein statistics, has to change, at a certain t, into a power. This could explain the break in t observed in the differential cross sections of certain reactions. It also follows that asymptotically the s and t dependences of the cross sections are of the same form, i.e. at large s a power-like decrease of cross sections with s, is matched at large t, by a power-like decrease of cross sections with t.

LIST OF PUBLICATIONS

BY

RICHARD M. WEINER

SELECTED LIST OF ARTICLES IN SCIENTIFIC JOURNALS

Influence of Argon on the vapour spectra of I, with I. Agarbiceanu, C. Ghiţa and V. Ţopa, *Comunicarile Academiei R.P.R.* 5 (1955) No. 1 (in Romanian).

The indeterminacy relation energy-time in non-relativistic quantum mechanics, *Studii siCercetari de Fizica* 6 (1955) 383 (in Romanian).

Nuclear isomerism and atomic spectra (1), *Studii şi Cercetari de Fizica* 7 (1956) 567.

Nuclear isomeric shift on spectral lines, *Nuovo Cimento* **4** (1956) 1587.

On the theoretical principles of the nuclear shell model, *Studii şi Cercetari de Fizica* 8 (1957) 359.

Charge distribution of excited isomeric nuclei and atomic spectra (The nuclear isomeric shift), *Phys. Rev.* **114** (1959) 256.

Nuclear isomerism and atomic spectra, *Soviet Physics JETP* 8 (1959) 196.

Effect of deformation of the nucleus on the electronic wave functions; application to beta decay, with Ch. Iussim, *Soviet Physics JETP* **11** (1960) 629.

Nuclear recoil in the decay of bound muons. A possible tool in muonic–atom studies, *Phys. Rev.* **138** (1965) B210.

Proposal to check C, P, and T conservation at large momentum transfer as a test of conventional quantum electrodynamics, *Phys. Rev.* **143** (1966) 1292.

Direct interaction mechanism in the ionization following beta decay, *Phys. Rev.* **144** (1966) 127.

Bosonic leptons, *Phys. Rev. Lett.* **18** (1967) 376.

SU(3) breaking by an isotriplet, with L. Banyai, N. Marinescu, V. Rittenberg, *Prog. Theor. Phys.* **37** (1967) 727.

Strange leptons, *Phys. Rev. Lett.* **20** (1968) 396.

Factorization quark model for large momentum transfer scattering (1), *Nucl. Phys. B* **6** (1968) 226.

The approximate C_γ invariance of the physical hadrons, *Rev. Roum. Phys.* **13** (1968) 31.

Do $\Delta S \neq 0$ semileptonic reactions really not conserve strangeness? *Rev. Roum. Phys.* **14** (1969) 1.

The antibaryon-baryon scattering problem, *Phys. Rev. D* **4** (1971) 813.

Natural versus unnatural parity exchange in forward scattering, *Phys. Rev. D* **4** (1971) 2174.

Maximum versus critical temperature in a superfluid approach to strong interactions, with A. Mann, *Lett. Nuovo Cimento* **2** (1971) 248.

Mesonic structure in the nucleon with K. Galloway and A. Mann, *Lett. Nuovo Cimento* **2** (1971) 635.

Dips in one-pion production processes from a statistical point of view, with K. Galloway and A. Mann, *Lett. Nuovo Cimento* **2** (1971) 1295.

Helicity amplitudes for antibaryon-baryon scattering (I), *Particles and Nuclei* **3** (1972) 140.

Helicity amplitudes for antibaryon-baryon scattering (II), *Particles and Nuclei* **3** (1972) 194.

Helicity amplitudes for antibaryon-baryon scattering (III), *Particles and Nuclei* **3** (1972) 264.

Derivation of the critical temperature for the superfluid phase of hadronic matter, with A. Mann, *Phys. Lett. B* **40** (1972) 383.

Superfluidity of hadronic matter, with A. Mann, *Nuovo Cimento A* **10** (1972) 625.

Statistical duality, *Lett. Nuovo Cimento* **8** (1973) 281.

Asymmetry in peripheral production processes, *Phys. Rev. Lett.* **32** (1974) 630.

Statistical approach to exchange reactions, *Acta Physics Austriaca* **40** (1974) 24.

SU(3) symmetry breaking and superfluidity of hadronic matter, with S. Eliezer, *Phys. Lett. B* **50** (1974) 463.

Preequilibrium and heat conduction in nuclear matter, with M. Weström, *Phys. Rev. Lett.* **34** (1975) 1523.

Propagation of heat in hadronic matter, *Phys. Rev. D* **13** (1976) 1363.

Phase transition of second and zero kind in high energy physics, with S. Eliezer, *Phys. Rev.* **13** (1976) 87.

Diffusion of heat in nuclear matter and preequilibrium phenomena, with M. Weström, *Nucl. Phys. A* **286** (1977) 282.

Energy distribution of secondaries in proton-nucleus collisions at very high energies, with N. Masuda, *Phys. Lett. B* **70** (1977) 77.

Possible evidence for coherence of hadronic fields from Bose-Einstein correlation experiments, with G. N. Fowler, *Phys. Lett. B* **70** (1977) 201.

Effects of classical fields in particle correlations, with G. N. Fowler, *Phys. Rev. D* **17** (1978) 3118.

Study of proton-nucleus collisions at high energies in the hydrodynamical model, with N. Masuda, *Phys. Rev. D* **18** (1978) 1515.

Hydrodynamical analysis of proton-nucleus collision data at 200 GeV, with N. Masuda, *Phys. Rev. D* **18** (1978) 1542.

Changes of symmetry at large transverse momenta, with S. Eliezer, *Phys. Rev. D* **18** (1978) 879.

Condensates and Bose-Einstein correlations, with G. N. Fowler and N. Stelte, *Nucl. Phys. A* **319** (1979) 349.

Evidence from very large transverse momenta of a change with temperature of velocity of sound in hadronic matter, with E. M. Friedlander, *Phys. Rev. Lett.* **43** (1979) 15.

Incompatibility of an exponential mass spectrum with the existence of quark matter, with G. N. Fowler, *Phys. Lett. B* **89** (1980) 394.

Cumulative effect and hot spots, with N. Stelte, *Phys. Lett. B* **103** (1981) 275.

Confinement and phase transitions, with G. N. Fowler and S. Raha, *Z. Phys. C* **9** (1981) 271.

Coherence and self-induced transparency in high energy hadronic collisions, with G. N. Fowler and E. M. Friedlander, *Phys. Lett. B* **104** (1981) 239.

Limiting fragmentation in high energy heavy ion reactions and preequilibrium, with S. Raha and N. Stelte, *Phys. Lett. B* **105** (1981) 411.

Solitons in nucleus-nucleus collisions near the speed of sound, with G. N. Fowler, S. Raha and N. Stelte, *Phys. Lett. B* **115** (1982) 286.

Drifting hot spots, with N. Stelte and M. Weström, *Nucl. Phys. A* **384** (1982) 130.

Centauros as a superfluorescence phenomenon, with G. N. Fowler and E. M. Friedlander, *Phys. Lett. B* **116** (1982) 203.

Are solitons already seen in heavy ion reactions? with S. Raha, *Phys. Rev. Lett.* **50** (1983) 407.

Particle multiplicities at the p-p collider and the hydrodynamical model, with E. M. Friedlander, *Phys. Rev. D* **28** (1983) 27.

Change of the vacuum outside a pion-condensed nucleus, with G. N. Fowler, S. Raha and J. Wheeler, *Phys. Rev. Lett.* **52** (1984) 891.

Limiting fragmentation and transparency in high energy heavy ion collisions, with R. Beckmann, S. Raha and N. Stelte, *Physica Scripta* **29** (1984) 197.

Effect of confinement on the velicity of sound in quark-gluon plasma, with M. Plümer and S. Raha, *Phys. Lett. B* **139** (1984) 198.

Spectator temperature as a signal for quark-matter formation, with S. Raha and J. Wheeler, *Phys. Rev. Lett.* **53** (1984) 138.

Inelasticity distribution and its implications for many particle production processes, with G. N. Fowler, E. M. Friedlander and M. Plümer *Phys. Lett. B* **145** (1984) 407.

Stability of density solitons formed in nuclear collisions, with S. Raha and K. Wehrberger, *Nucl. Phys. A* **433** (1984) 427.

Role of the equation of state in the hydrodynamical model, with K. Wehrberger, *Phys. Rev. D* **31** (1985) 222.

Stable vortex excitations in rotating nuclei, with G. N. Fowler and S. Raha, *Phys. Rev. C* **3** (1985) 1515.

Derivation of inelasticity distribution for hadron-hadron collisions in a gluon-gluon interaction picture, with G. N. Fowler and G. Wilk, *Phys. Rev. Lett.* **55** (1985) 173.

Production of pions in a coherent state in heavy-ion collisions, with G. N. Fowler, *Phys. Rev. Lett.* **55** (1985) 1373.

Effect of the nuclear surface on a propagating density pulse, with E. F. Hefter and S. Raha, *Phys. Rev. C* **32** (1985) 2201.

High energy behaviour of multiparticle distributions and their bounds with G. N. Fowler, E. M. Friedlander and G. Wilk, *Phys. Rev. Lett.* **56** (1986) 14.

Possible manifestation of quark-gluon plasma in multiplicity distributions from high energy reactions, with G. N. Fowler, E. M. Friedlander and G. Wilk, *Phys. Rev. Lett.* **57** (1986) 2119.

Approach to chaos in high energy particle reactions, with P. Carruthers and E. M. Friedlander, *Physica D* **23** (1986) 138.

Expansion of the early universe and the equation of state, with U. Ornik, *Phys. Rev. D* **36** (1987) 1263.

Inelasticity and leading particle effect: Momentum and mass distribution of the central fireball in high-energy hadronic interactions, with G. N. Fowler, A. Vourdas and G. Wilk, *Phys. Rev. D* **36** (1987) 870.

Photon counting distribution in squeezed states, with A. Vourdas, *Phys. Rev. A* **36** (1987) 5866.

Multiplicity distributions in hadronic and heavy ion reactions, *Z. Phys. C* **38** (1988) 199.

Initial energy density distribution in high energy heavy ion reactions, with G. N. Fowler, F. S. Navara, M. Plümer, A. Vourdas and G. Wilk, *Phys. Lett. B* **214** (1988) 657.

Multiparticle correlations from transverse-energy distributions, with M. Plümer, *Phys. Rev. D* **37** (1988) 3136.

Rapidity scaling of multiplicity distributions in a quantum-statistical approach, with G. N. Fowler, E. M. Friedlander, F. W. Pottag, J. Wheeler and G. Wilk, *Phys. Rev. D* **37** (1988) 3127.

Entropy scaling in high energy collisions. Chaos and coherence, with P. A. Carruthers, M. Plümer and S. Raha, *Phys. Lett. B* **212** (1988) 369.

Multiplicity distributions and Bose-Einstein correlations in high-energy multiparticle production in the presence of squeezed coherent states, with A. Vourdas, *Phys. Rev. D* **38** (1988) 2209.

Subthreshold multiple pion production in nucleus-nucleus collisions, with E. M. Friedlander and F. S. Navara, *Modern Phys. Lett. A* **3** (1988) 1461.

Multiplicity distributions in e^+-e^- and μ-p reactions and their rapidity dependence from a quantum-statistical point of view, with E. M. Friedlander and F. W. Pottag, *J. Phys. G: Nucl. Part. Phys.* **15** (1989) 431.

Multiplicity fluctuations in finite rapidity windows. Intermittency or quantum statistical correlation? with P. A. Carruthers, E. M. Friedlander and C. C. Shih, *Phys. Lett. B* **222** (1989) 487.

Forward-backward multiplicity distribution in a two-component model, with M. Biyajima, G. N. Fowler, N. Suzuki and G. Wilk, *Z. Phys. C* **44** (1989) 199.

Interacting gluon model for hadron-nucleus and nucleus-nucleus collisions in the central rapidity region, with G. N. Fowler, F. S. Navara, M. Plümer, A. Vourdas and G. Wilk, *Phys. Rev. C* **40** (1989) 1219.

Hadron interferometry revisited, *Phys. Lett. B* **232** (1989) 278; *ibid.* (Erratum) *B* **242** (1990) 547.

High-energy heavy-ion reactions and relativistic hydrodynamics in three dimensions, with U. Ornik and F. W. Pottag, *Phys. Rev. Lett.* **63** (1989) 2641.

Energy dependence of multiplicity distributions in the quantum statistical approach, with G. N. Fowler, E. M. Friedlander, F. S. Navarra, F. W. Pottag and G. Wilk, *J. Phys. G: Part. Phys.* **16** (1990) 1439.

Space-time aspects of Bose-Einstein correlations and quantum statistics, with I. V. Andreev, *Phys. Lett. B* **253** (1991) 416.

Multiplicity dependence of Bose-Einstein correlations and quantum statistics, with G. N. Fowler, E. M. Friedlander, X. C. He and C. C. Shih, *Phys. Lett. B* **253** (1991) 421.

Derivation of the two-component model for multiplicity distributions from a stochatic branching mechsnism, with M. Biyajima, G. N. Fowler, E. M. Friedlander, C. C. Shih, N. Suzuki and G. Wilk, *Phys. Rev. D* **43** (1991) 43.

Soft pion enhancement as a consequence of hydrodynamical expansion and final state interactions, with U. Ornik, *Phys. Lett. B* **263** (1991) 503.

Multiplicity distributions, transverse momenta, and nonstationary effects, with E. M. Friedlander, X. C. He and C. C. Shih, *Phys. Rev. D* **44** (1991) 1396.

Surprises from Bose-Einstein correlations, with I. V. Andreev and M. Plümer, *Phys. Rev. Lett.* **67** (1991) 3475.

Correlation length versus radius in Bose-Einstein correlations, with I. V. Andreev, *Nucl. Phys. A* **525** (1991) 527c.

Limited growth of the number of sources and decrease of inelasticity in high-energy reactions, with G. Wilk and Z. Wlodarczyk, *Phys. Rev. D* **45** (1992) 2308.

Energy dependence of inelasticity from high-energy cosmic ray reactions, with Yu. M. Shabelski, G. Wilk and Z. Wlodarczyk, *J. Phys. G* **18** (1992) 1281.

Evidence for quantum statistical coherence from experimental data on higher order Bose-Einstein correlations, with M. Plümer and L. Razumov, *Phys. Lett. B* **286** (1992) 335.

Transverse expansion and effective radii of sources in high energy collisions, with B. Schlei, U. Ornik and M. Plümer, *Phys. Lett. B* **293** (1992) 275.

Relativistic hydrodynamics of partially stopped baryonic matter, with J. Bolz and U. Ornik, *Phys. Rev. C* **46** (1992) 2047.

Interferometry of pions and kaons in high energy collisions, with J. Bolz, U. Ornik, M. Plümer and B. R. Schlei, *Phys. Lett. B* **300** (1993) 404.

Resonance decays and partial coherence in Bose-Einstein correlations, with J. Bolz, U. Ornik, M. Plümer and B. R. Schlei, *Phys. Rev. D* **47** (1993) 3860.

Relationship between Bose-Einstein correlations and intermittency revisited, with I. V. Andreev, M. Plümer and B. R. Schlei, *Phys. Lett. B* **316** (1993) 583.

Quantum-statistical space-time approach to Bose-Einstein correlations and multiplicity distributions, with I. V. Andreev and M. Plümer, *Int. J. Mod. Phys. A* **8** (1993) 4577.

Dynamical properties of heavy-ion collisions from photon intensity interferometry, with L. V. Razumov, *Phys. Lett. B* **319** (1993) 431.

Evidence for long range fluctuations in Bose-Einstein correlation measurements, with I. V. Andreev, M. Plümer and B. R. Schlei, *Phys. Lett. B* **321** (1994) 277.

Bose-Einstein correlations in the invariant q^2 variable and intermittency, with I. V. Andreev, M. Plümer and B. R. Schlei, *Phys. Lett. D* **49** (1994) 1217.

Bounds for Bose–Einstein correlation functions, with M. Plümer and I. V. Razumov. *Phys. Rev. D* **49** (1994) 49.

Photon interferometry of quark-gluon dynamics re-examined, with A. Timmermann, M. Plümer and L. Razumov, *Phys. Rev. C* **50** (1994) 3060.

Thermal photon production in heavy-ion collisions, with N. Arbex, U. Ornik, M. Plümer and A. Timmermann, *Phys. Lett. B* **345** (1995) 307.

Quantum field theory of Bose-Einstein correlations, with L. Razumov, *Phys. Lett. B* **348** (1995) 133.

Production of a chaotic squeezed state from a "pion liquid" and overbunching of identical pion correlations, with I. V. Andreev, *Phys. Lett. B* **373** (1996) 159.

Hydrodynamical analysis of single inclusive spectra and Bose-Einstein correlations for Pb+Pb at 160 AGeV, with B. R. Schlei, U. Ornik, M. Plümer and D. Strottman, *Phys. Lett. B* **376** (1996) 212.

Hydrodynamical analysis of symmetric nucleus-nucleus collisions near 200A GeV, with U. Ornik, M. Pluemer, B. R. Schlei and D. Strottman, *Phys. Rev. C* **54** (1996) 1381.

Probing the equation of state in Au+Au at 11 GeV/nucleon with (3+1)-dimensional hydrodynamics, with N. Arbex, U. Ornik and M. Pluemer, *Phys. Rev. C* **55** (1997) 860.

π^--π^+ratio in heavy ion collisions: Coulomb effect or chemical equilibration?, with N. Arbex, U. Ornik, M. Pluemer and B. R. Schlei, *Phys. Lett. B* **391** (1997) 465.

Effects of Bose condensates on single inclusive spectra and Bose-Einstein correlations, with U. Onik, M. Pluemer, B. R. Schlei and D. Strottman, *Phys. Rev. C* **56** (1997) 412.

Higher order Bose-Einstein Correlations test the Gaussian density matrix and the space-time approach, with N. Arbex and M. Plümer, *Phys. Lett. B* **438** (1998) 193.

Boson interferometry in high-energy physics, *Phys. Rep.* **327** (2000) 249.

Surprises from the search for quark-gluon plasma? When was quark-gluon plasma seen?, *Int. J. Mod. Phys. E* **15** (2006) 37.

BOOKS AUTHORED

Introduction to Bose-Einstein Correlations and Subatomic Interferometry, John Wiley and Sons, 2000.

Das Miniatom-Projekt, Verlag LiteraturWissenschaft. de (TransMIT), 2006 (in German).

BOOKS EDITED

Local Equilibrium in Strong Interaction Physics, co-ed. D. K. Scott, World Scientific 1985.

Correlations and Multiparticle Production (CAMP), co-eds. M. Plümer, S. Raha, World Scientific 1991.

Bose-Einstein Correlations in Particle and Nuclear Physics, A Collection of Reprints, John Wiley and Sons, 1997.

SELECTED LIST OF INVITED PAPERS AND CONTRIBUTIONS PUBLISHED IN BOOKS AND PROCEEDINGS OF SCIENTIFIC MEETINGS

Changes of symmetries in neutrino reactions, with S. Eliezer, *Proceedings International Neutrino Conference*, Aachen, eds. H. Faissner, H. Reithler and P. Zerwas, Vieweg, 1976, p. 661.

Possible evidence for coherence of hadronic fields from Bose–Einstein correlation experiments, *8th Meeting on Multiparticle Dynamics*, Kaysersberg, 1977, p. A277.

What can be learned from boson correlations in p-p and e^+-e^- reactions?, with G. N. Fowler, *IVth European Symposium on Antiprotons,* Strasbourg,1978, p. 319.

Geometry and dynamics in the hot spot model, with N. Stelte, *Lecture Notes in Physics Vol. 117,* Springer, 1979, p. 161.

How free is the quark-gluom plasma?, with M. Plümer and S. Raha, *Quark Matter 1983, Nucl. Phys. A* **418** (1984) 549c.

Local excitations (hot spots) in nuclear matter, *Coincident Particle Emission from Continuum States in Nuclei*, eds. H. Machner and P. Jahn, World Scientific, 1984, p. 391.

Mean free path in strong interactions, *1st International Workshop on Local Equilibrium in Strong Interaction Physics* (*LESIP I*), eds. D. K. Scott and R. M. Weiner, World Scientific, 1985, p. 419.

The Landau Hydrodynamical Model (Round Table Discussion), *ibid*, p. 447.

Will multiplicities go on broadening for ever? *Multiparticle Dynamics*, ed. J. Grunhaus, World Scientific, 1985, pp. 331–343.

Quark-gluon plasma and the Landau hydrodynamical model, *Hadronic Matter Under Extreme Conditions,* Kiev, eds. G. M. Zinoviev and V. P. Shelest, Naukova Dumka, 1986, p. 261.

Multiparticle phenomenology as a synthesis of microscopic and statistical concepts, *Hadronic Matter in Collision* (*LESIP II*), Santa Fe, eds. P. Carruthers and D. Strottman, World Scientific, 1986, p. 106.

Hot spots — A "common market" of nuclear and elementary particle Physics, *International Workshop on Reaction Models for*

Continuous Spectra of Light Particles, ed. J. Ernst, Gesellschaft für Physikalische Forschung und Naturwissenschaftlich — Technische Weiterbildung, Institut für Strahlen- und Kernphysik Universität Bonn, 1986, p. 129.

Transverse energy and multiplicity distribution in high energy physics at future accelerators, *10th Warsaw Symposium on Elementary Particle Physics*, ed. Z. Ajduc, 1987, p. 103.

What do we learn from the investigation of multiparticle production processes?, *Shandong Workshop on Multiparticle Production,* eds. R. Hwa and X. Qu, World Scientific, 1988, p. 423.

Application of the methods of quantum optics to multihadron production, with G. N. Fowler, *Hadronic Multiparticle Production*, ed. P. Carruthers, World Scientific, 1988, p. 481.

Coherent states in high energy physics revisited, *Hadronic Matter in Collision* (*LESIP III*), Tucson, eds. P. Carruthers and J. Rafelski, World Scientific, 1988, p. 175.

Rapidity and transverse momentum distributions of secondaries produced in O+Au collissions at 200 A.GeV from a 2+1 dimensional solution of relativistic hydrodynamics, with F. W. Pottag and U. Ornik, *ibid*., p. 310.

From two particle correlation to multiplicity distributions multiparticle production, *Multiparticle Production*, Perugia, eds. R. Hwa, G. Pancheri and Y. Srivastava, World Scientific, 1988, p. 231.

Equation of state in nuclear, hadronic and quark matter, *International School of Physics Enrico Fermi, Varena, High Pressure Equations of State*, eds. S. Eliezer and R. A. Ricci, North Holland, 1989, p. 535.

Remarks on Bose-Einstein correlations, *International Workshop on Particle Correlations and Interferometry in Nuclear Collisions*, Nantes, ed. D. Ardouin, World Scientific, 1990, p. 459.

Soft pions and geometry in Bose-Einstein correlations, with I. V. Andreev, *Correlations and Multiparticle Production* (*CAMP — LESIP IV*), Marburg, eds. M. Plümer, S. Raha and R. M. Weiner, World Scientific, 1991, p. 80.

Multiplicity distributions from calorimetric measurements of transverse energy, with S. Raha, G. Wilk and Z. Wlodarczyk, *ibid.*, p. 283.

The number of sources in multiparticle production, with G. Wilk and Z. Wlodarczyk, *ibid.*, p. 388.

Bose-Einstein correlations and quantum statistics, *ibid.*, p. 434.

A strategy for multiparticle dynamics, *Fluctuations and Fractal Structure*, Ringberg Castle, eds. R. C. Hwa, W. Ochs and N. Schmitz, World Scientific, 1991, p. 321.

Soft pions, Bose-Einstein correlations, squeezed states and intermittency, *High Energy Nuclear Collisions and Quark Gluon Plasma*, Kyoto, eds. M. Biyajima, H. Enyo, T. Kunihiro and O. Miyamura, World Scientific, 1992, p. 253.

Coherence and chaos in the interacting gluon model, with G. Wilk and Z. Wlodarczyk, *Particles and fields, Proceedings of the First German-Polish Symposium on Particles and Fields*, Rydzyna Castle, Poland, 28 April–1 May 1992, eds. H. D. Doebner, M. Pawlowski and R. Raczka, World Scientific, 1994, p. 193.

Hydrodynamical beam jets in high energy hadronic collisions, with U. Ornik and G. Wilk, *Quark Matter 1993, Nucl. Phys. A* **566** (1994) 469c.

Modelling coherence and chaos in high-energy collisions, with G. Wilk and Z. Wlodarczyk, *Proceedings of the Workshop on Pre-Equilibrium Parton Dynamics in Heavy Ion Collisions*, LBL Berkeley, August 23–September 3, 1993, ed. X. N. Wang, LBL-Report 34831 (1993), p. 305.

Correlations and strong interactions, *Hot Hadronic Matter*, eds. J. Letessier, H. Gutbrod and J. Rafelski, Plenum Press, 1995, p. 323.

Bose–Einstein correlations and quantum field theory, *Acta Phys. Pol. B* **27** (1996) 2637.

Change with energy of source geometry as seen in hadron interferometry *XXIV Multiparticle Dynamics, Vietri sul Mare,* eds. A. Giovannini, S. Lapin and R. Ugoccioni, World Scientific, 1994, p. 127.

Explosion of a hadronic fireball, *XXV Multiparticle Dynamics*, eds. D. Bruncko, L. Sandor and J. Urban, World Scientific, 1995, p. 644.

Bose-Einstein correlations: A research program for the 21st century. *Correlations and Fluctuations*, Nijmegen, ed. W. Kittel, World Scientific, 1997, p. 1.

Photon versus hadron interferometry, *Measuring the Size of Things in the Universe CRIS98 Meeting*, eds. S. Costa *et al.*, Catania, World Scientific, 1999, p. 178.

Impressionism and surrealism in multiparticle dynamics, *Correlations 98*, Matrahaza, eds. T. Csorgo *et al.*, World Scientific, 1999, p. 245.

Notes and References

1. *Die Zeit*, October 24, 1997; *ibid.* April 24, 2003.
2. German–French TV channel Arte 1999, documentary by Koepp.
3. "A Poet at War with His Language," *The New York Times*, December 31, 2000.
4. Hulubei predicted and observed the multiple Compton scattering effect experimentally. However, his claimed discovery, a new element, moldavium, was never confirmed.
5. I. Agărbiceanu, C. Ghiţa, V. Ţopa and R. M. Weiner, *Comunicarile Academiei R.P.R.* **5**(1) (1955) (in Romanian).
6. R. M. Weiner, *Studii şi Cercetari de Fizica* **6** (1955) 383 (in Romanian).
7. In analogy to the fine structure of atomic spectra, which is due to the electron spin, the hyperfine structure is due to the effect of the nuclear magnetic dipole moment associated with the nuclear spin and of the electric quadrupole moment.
8. U. Gonser (ed.), *Mössbauer Spectroscopy*, Springer 1975, p. 2.
9. R. M. Weiner, *Nuovo Cimento* **4** (1956) 1587.
10. R. M. Weiner, *Phys. Rev.* **114** (1959) 256.
11. D. A. Shirley, Nuclear applications of isomeric shifts, *Proc. Int. Conf. on the Mössbauer Effect*, Saclay, 1961, eds. D. H. Compton and A. H. Schoen, John Wiley & Sons, New York, p. 258.
12. A. Melissinos and S. Davis, *Phys. Rev.* **115** (1959) 130.
13. Furthermore, deformation and compressional effects, which could be neglected for In, were more important in Hg.
14. J. Lardinois, *Nucl. Phys.* **15** (1960) 522.
15. H. Frauenfelder, *The Mössbauer Effect*, W. A. Benjamin, Inc., 1964, p. 54.
16. R. Mössbauer, *Z. Phys.* **151** (1958) 124.
17. O. C. Kistner and A. W. Sunyar, *Phys. Rev. Lett.* **4** (1960) 412.
18. G. M. Kalvius, Evaluation of isomer shift, in *Hyperfine Interactions in Excited Nuclei*, eds. G. Goldring and R. Kalish, Gordon and Breach Science Publisher, N.Y. 1971, p. 523.
19. A. Steudel, private communication, March 3, 1978.
20. G. K. Shenoy, private communication, January 23, 1978.

21. F. Wagner, private communication, December 12, 1977.
22. S. L. Ruby, in *Mössbauer Isomer Shifts*, eds. G. K. Shenoy and F. E. Wagner, North Holland Publishing Company, 1978.
23. C. S. Wu, private communication, February 3, 1978.
24. Nuclear isomerism and atomic spectra, *Sov. Phys. JETP* **8**, 196.
25. *Sov. Phys. JETP* **11** (1960) 629.
26. Among others, Wold (*Nucl. Phys. A* **97** (1967) 620) confirmed our results and Maltese (*Phys. Rev.* **173** (1968) 1165) applied them in the interpretation of experimental data. Even in 1975 it was referred to (by Bergkvist).
27. I am aware of at least six papers in this category published in the period 1965–1968. Between 1968 and 1969, when I left Romania, I was allowed to publish indicating the affiliation with the University Computing Center.
28. *Phys. Rev. B* **138** (1965) 210.
29. The parity symmetry P implies that the simultaneous reflection of the coordinates x, y, z around the origin leaves the physics laws unchanged; the charge conjugation symmetry C means that the same laws govern matter and anti-matter; the PC symmetry implies time (T) reversibility, because the product PCT is (most probably) conserved.
30. R. B. Blumenthal *et al.*, Deviation from simple Quantum Electrodynamics, *Phys. Rev. Lett.* **14** (1965) 660.
31. R. M. Weiner, Proposal to check C, P, and T conservation at large momentum transfer as a test of conventional quantum electrodynamics, *Phys. Rev.* **143** (1966) 1292.
33. Leptons are particles that interact only weakly and electromagentically, while hadrons (nucleons and mesons) are particles that also interact strongly. The pion is the lightest hadron and it is a meson; the muon is a charged lepton, quite similar to the electron, albeit more massive, and the neutrino is a neutral lepton. Mesons are bosons i.e. particles with integer spin; those observed so far do not have intrinsic quantum numbers. Leptons are fermions, i.e. particles with non-integer spin (1/2) and which have a specific internal quantum number.
34. R. M. Weiner, Bosonic leptons, *Phys. Rev. Lett.* **18** (1967) 376.
35. R. M. Weiner, Strange leptons, *Phys. Rev. Lett.* **20** (1968) 396.
36. Cf. e.g. E. F. Beall, On strange leptons, *Phys. Rev. Lett.* **20** (1968) 947; K. H. Young, J. A. Helland and M. J. Longo, Experimental evidence against the existence of strange leptons, *Phys. Rev. Lett.* **21** (1968) 254; E. Fishbach, T. Kirsten and O. A. Schaeffer, Experimental limits on the existence of anomalous electrons, *Phys. Rev. Lett.* **20** (1968) 1012; K. W. Rothe and A. M. Wolsky, Are there heavy leptons?, *Nucl. Phys. B* **10** (1969) 241; J. H. Munsee and F. Reines, Search for a neutral weak interaction via ν_e dissociation of deuterons, *Phys. Rev.* **177** (1969) 2002; D. Ljung and D. Cline, Experimental study of the rare K^+ decay modes, *Phys. Rev. D* **8** (1973) 1307; G. R. Kalbfleisch and E. C. Fowler, Color of the quark 'muon', *Nuovo Cimento A* **19** (1974) 173; S. Pakvasa and K. Tennakone, Neutrino spectrum and the origin of the Cabibbo angle, *Lett. Nuovo Cimento* **9** (1974) 565.

37. The electromagnetic interactions of strange charged leptons would have to be different from those of non-strange ones.
38. Cf. e.g. Glashow's paper "Neutrino Identity," *J. Phys.* **32** (1971) C3.
39. Harcourt, Brace, New York, 1951.
40. R. Gellately, *Consent and Coercion in Nazi Germany*, Oxford University Press 2002, quoted in *Die Zeit*, Literatur, March 2002, p. 58.
41. Cf. e.g. A. Bortschagowski, *Orden für einen Mord*, Propyläen Verlag, Berlin 1997; F. J. Raddatz, *Die Zeit*, October 17, 1997, p. 35.
42. The name was misprinted, which did not prevent the CERN postmaster from finding me.
43. G. N. Fowler and R. M. Weiner, *Phys. Lett. B* **89** (1980) 394.
44. In the late seventies, the idea of quark deconfinement led to a revision of the statistical bootstrap; Hagedorn and Rafelski, by introducing the finite extension of hadrons, generalized it, and this made possible, within this model, a phase transition from hadronic matter to quark matter.
45. P. Carruthers, *Annals New York Acad. Sci.* **229** (1974) 91; cf. also P. Carruthers and M. Duong-Van, *Phys. Rev. D* **8** (1973) 859.
46. D. A. Kirzhnits and A. D. Linde, *Phys. Lett. B* **42** (1972) 471.
47. Phonons are the quanta through which sound propagates. Like their cousins, the photons, which are the light quanta, they are massless.
48. A. Mann and R. M. Weiner, *Lett. Nuovo Cimento* **2** (1971) 248.
49. K. F. Galloway, A. Mann and R. M. Weiner, *Lett. Nuovo Cimento* **2** (1971) 625.
50. This effective temperature found referred only to the cloud, i.e. a subsystem of the nucleon.
51. This point led initially to misunderstandings. It was discussed at greater length, much later, by Van Hove in *Z. Phys. C* **21** (1983) 93.
52. A. Mann and R. M. Weiner, *Nuovo Cimento A* **10** (1972) 625.
53. T. T. Chou and C. N. Yang, *Phys. Rev. D* **4** (1971) 2005).
54. A. Mann and R. M. Weiner, *Phys. Lett. B* **40** (1972) 383.
55. Phase transitions are classified in terms of the analytical properties of certain thermodynamical variables like free energy, pressure, etc. Phase transitions of first kind involve discontinuities of first order in these variables while transitions of second or higher order refer to discontinuities of second or higher order.
56. Spontaneously broken symmetries occur in systems the symmetry of which is higher than that of their lowest energy state, the vacuum.
57. S. Eliezer and R. M. Weiner, SU(3) symmetry breaking and superfluidity of hadronic matter, *Phys. Lett. B* **50** (1974) 463.
58. Color is a quantum property characterizing quarks. Besides color, quarks have, like leptons, also flavor quantum numbers. The symmetries associated with these quantum numbers are mathematically described by groups. In particle physics, "special unitary groups" — denoted by SU(N), where N is the dimension of the group — play a particularly important role.
59. S. Eliezer and R. M. Weiner, *Phys. Rev. D* **13** (1976) 87.

60. Chiral symmetry is a symmetry involving the relative orientation of spin and momentum of a particle. It plays an essential role in the standard model of strong and electroweak interactions.
61. Cf. the review by A. Linde, Phase transitions in gauge-theories and cosmology, *Rep. Prog. Phys.* **42** (1979) 389.
62. The 1972 paper by Kirshnitz and Linde of Ref. 46 refers to weak interactions only and its applications are limited to cosmology.
63. G. N. Fowler, S. Raha and R. M. Weiner, *Z. Phys. C* **9** (1981) 271.
64. M. Plümer, S. Raha and R. M. Weiner, *Nucl. Phys. A* **418** (1984) 549c; M. Plümer, S. Raha and R. M. Weiner, *Phys. Lett. B* **139** (1984) 198.
65. G. Chapline, *Phys. Rev. D* **11** (1975) 156.
66. In an erratum published a few months later (*Phys. Rev. D* **12** (1975) 1515), Chapline writes: "The article failed to note that that hadronic matter might have superfluid properties had been previously discussed in a series of articles by A. Mann and R. M. Weiner. In particular, the possibility that the mesons inside a nucleon behave as a superfluid was reached for the first time by Mann and Weiner (*Lett. Nuovo Cimento* **2** (1971) 248), starting from the observation that the dispersion relation of the pion cloud of nucleon satisfies the Landau criterion of superfluidity. Some consequences of this idea for the strong interactions were discussed by Mann and Weiner in *Nuovo Cimento A* **10** (1972) 624; *Phys. Lett. B* **40** (1972) 383 and by S. Eliezer and R. M. Weiner in *Phys. Lett. B* **50** (1974) 463.
67. D. F. Goble, *Ann. Phys.* **90** (1975) 295.
68. Y. Nambu and G. Jona-Lasinio, *Phys. Rev.* **122** (1961) 345; *ibid*, **124** (1961) 246.
69. J. Chela-Flores, *Phys. Rev. D* **18** (1978) 2632; *Nuovo Cimento A* **58** (1980) 67.
70. J. Chela-Flores, *Nuovo Cimento A* **68** (1982) 308.
71. J. Chela-Flores, *Phys. Rev. D* **29** (1984) 1339.
72. J. Chela-Flores and V. Varela, *Phys. Rev. D* **27** (1983) 1248.
73. M. M. Islam, R. J. Luddy and A. V. Prokudin, *Mod. Phys. Lett. A* **18** (2003) 743 [hep-ph/0210437]; *Mod. Phys. Lett. A* **18** (2003) 743 and references therein.
74. G. N. Fowler and R. M. Weiner, *Phys. Lett. B* **89** (1980) 394.
75. A. Salam and J. Strathdee, *Nature* **252** (1974) 569.
76. S. Eliezer and R. M. Weiner, *Phys. Rev. D* **18** (1978) 879.
77. Indeed, a few years later the mass of the W and of the Z turned out to be 80 GeV and 91 GeV, respectively.
78. E. V. Filipova *et al.*, *Astron. Lett.* **31** (2005) 729.
79. "Above" and "below" have to be defined, of course, for each event separately.
80. R. M. Weiner, *Phys. Rev. Lett.* **32** (1974) 630.
81. R. M. Weiner, *Phys. Rev. D* **13** (1976) 1363.
82. J. Goldberg, *Phys. Rev. Lett.* **43** (1979) 250.
83. The issue of hot spots in deep inelastic scattering was raised again, among others in 1990, and again at DESY, in connection with the saturation of

quark and gluon densities to be expected at the DESY accelerator HERA (*CERN Courier*, July/August 1990, p. 23).
84. G. A. Miller, *Phys. Rev. C* **68** (2003) 022201.
85. R. M. Weiner and M. Weström, *Phys. Rev. Lett.* **34** (1975) 1523; *Nucl. Phys. A* **286** (1977) 282.
86. T. Sugitate *et al.*, *Nucl. Phys. A* **388** (1982) 402.
87. W. Trautmann *et al.*, *Phys. Rev. Lett.* **53** (1984) 1630.
88. M. Gyulassy *et al.*, *Nucl. Phys. A* **613** (1997) 397.
89. H. Bethe, *Phys. Rev. Abstracts* **53** (1938) 675.
90. Cf. e.g. J. Knab, *Die Zeit*, November 10, 2005.
91. R. Gellately, *Consent and Coercion in Nazi Germany*, Oxford University Press 2002, quoted by V. Ullrich in *Die Zeit*, March 2002.
92. B. Greiner, *Die Zeit*, October 8, 1993.
93. E. Klee, *Die Zeit*, September 25, 2003.
94. Cf. e.g. K. Adam, "Messen mit zweierlei Maß," *Die Welt*, December 11, 2003.
95. Cf. e.g. H. Jensen, *Die Zeit*, April 24, 1992.
96. *Ibid.*
97. Interview of Weizsäcker in *Die Zeit*, June 26, 1992.
98. B. Greiner, *Die Zeit*, October 8, 1993.
99. Quoted by B. Greiner, *Die Zeit*, October 8, 1993.
100. Cf, N. Bohr Archives, www.nbi.dk/NBA.
101. *Die Zeit*, April 24, 1992.
102. J. Trenkner, *Die Zeit*, February 6, 2003.
103. *Frankfurter Allgemeine Zeitung*, August 11, 1994.
104. A. Paul, *Die Zeit*, January 5, 2006.
105. One example is the case of Franz Nüßlein, a condemned war criminal, reintegrated in 1955 into the diplomatic corps (cf. *Die Zeit*, January 26, 2006, p. 92).
106. Woe to the vanquished.
107. Bremsstrahlung, as its name suggests, is the emission of particles due to the "braking" of the projectile in the presence of another particle. The difference between the weak and strong coupling regime manifests itself through the fact that in electrodynamics multiple production occurs only at extremely high energies, while in mesodynamics it is expected to happen also at lower energies (W. Heisenberg, *Z. Phys.* **101** (1936) 569).
108. W. Heisenberg *Z. Phys.* **101** (1936) 533; *ibid.* **113** (1939) 61.
109. As a matter of fact, this statement has to be qualified: in the 1980s, that is five decades after Heisenberg's analogy with turbulence, a similar idea, this time in terms of intermittency, was tried (cf. Chap. 23).
110. E. Fermi, *Prog. Theor. Phys.* **5** (1950) 570.
111. L. Landau, *Izv. Akad. Nauk (USSR) Ser. Fiz.* **17** (1953) 51; translated in *Collected Papers*, ed. D. Ter Haar (Pergamon, Oxford, 1965), p. 569.
112. E. L. Feinberg, *Phys. Rep.* **5** (1972) 237.
113. E. M. Friedlander, *Lett. Nuovo Cimento* **9** (1974) 349.
114. Shock waves are specific discontinuities in the propagation of waves.

115. Cf. also Chaps. 27 and 28.
116. The one-dimensional approximation is justified by the idea that the initially compressed system has at first to decompress before emission takes place. For the same reason, this approximation was also used by Landau for proton–proton reactions.
117. At the end of the 1970s, with the realization of the possible existence of a phase transition from hadronic matter to quark–gluon plasma, the situation became even more complicated. This issue will be addressed later.
118. N. Masuda and R. M. Weiner, *Phys. Rev. D* **18** (1978) 1515, 1542.
119. By boundary conditions, I mean here among other things the freeze-out mechanism.
120. For a qualification of this statement due to the remnant effect of confinement cf. the discussion below.
121. E. M. Friedlander and R. M. Weiner, *Phys. Rev.* **28** (1983) 2903.
122. G. N. Fowler, E. M. Friedlander, M. Plümer and R. M. Weiner, *Phys. Lett. B* **145** (1984) 407; G. N. Fowler, R. M. Weiner and G. Wilk, *Phys. Rev. Lett.* **55** (1985) 173; G. N. Fowler, A. Vourdas, R. M. Weiner and G. Wilk, *Phys. Rev. D* **35** (1987) 870; G. N. Fowler, F. S. Navara, M. Plümer, A. Vourdas and G. Wilk, *Phys. Lett. B* **214** (1988) 657; G. N. Fowler, F. S. Navara, M. Plümer, A. Vourdas and G. Wilk, *Phys. Rev. C* **40** (1989) 1219; G. N. Fowler, E. M. Friedlander, F. S. Navarra, W. Pottag and G. Wilk, *J. Phys. G: Part. Phys.* **16** (1990) 1439; Yu. M. Shabelski, R. M. Weiner, G. Wilk and Z. Wlodarczyk, *J. Phys. G* **18** (1992) 1281.
123. Inelasticity K is not the only physical quantity in the initial conditions, which led to new developments. The issue of Lorentz contraction has also been a matter of debate, in particular at the 1984 LESIP I meeting, since the value of the inelasticity can hardly be separated from the amount of Lorentz contraction: both these quantities enter the energy density, which is the quantity that determines the initial temperature and the hydrodynamical evolution of the system. It is interesting to mention that for a quantum system the value of the Lorentz factor is determined by the form of the Hamiltonian and does not necessarily coincide with the classical value (Al. Müller, *Phys. Rev. D* **2** (1970) 2241). I also discussed this issue in a review (R. M. Weiner, in *Hadronic Matter Under Extreme Conditions*, Kiev, Naukove Dumka, 1986, eds. G. M. Zinoviev and V. P. Shelest, p. 261), where I suggested that in the Landau model there might be no Lorentz contraction at all and that we are rather faced with a quantum system of energy $K^j s$ the (longitudinal) dimensions of which are given by the de Broglie wavelength $1/(K^j s)$. It seems that the last word on this issue has not yet been said.
124. Entropy is the measure of a system's energy that is unavailable for work. Since work is obtained from order, the amount of entropy is also a measure of the disorder, or randomness, of the system.
125. The proceedings of these meetings are published in: *Local Equilibrium in Strong Interactions Physics* (*LESIP I*), eds. D. K. Scott and R. M. Weiner, World Scientific 1985; *Hadronic Matter in Collision* (*LESIP II*), eds.

P. Carruthers and D. Strottman, World Scientific 1986; *Hadronic Matter in Collision 1988 (LESIP III)*, eds. P. Carruthers and J. Rafelski, World Scientific 1988; *Correlations and Multiparticle Production (LESI IV)*, eds. M. Plümer, S. Raha and R. M. Weiner, World Scientific 1991.

126. A. Unsöld, *Phys. Blätter* **36** (1980) 337.
127. The actual date of the colloquium was April 25, 1980.
128. In a letter of March 16, 1981 addressed to the Nobel laureate, Kastler Hanle says that he was afraid of being considered "biased."
129. Translation from German and explanations by R. W.
130. Unsöld quotes in his article from Romain Rolland's diary which mentions that in 1915 Einstein dreamed of a divided Germany: Southern Germany and Austria as opposed to Prussia.
131. Unsöld alludes here to a law of the Federal Republic of Germany forbidding members of the Communist party to be civil servants.
132. Hagedorn made part of Rommel's Afrika Korp before being taken prisoner by the British (R. W).
133. This bracket appears in Hagedorn's letter as a footnote.
134. Para-Military sport groups with Nazi background (R. W.).
135. *Phys. Blätter* **37** (1981) 85.
136. *Nature* **290** (1981) 535
137. "Einstein en procès: les vieux démons ne sont pas morts" (Einstein's trial: The old demons are not dead), *La Recherche* **12** (1981) 668.
138. "Unsöld on Einstein," *Discover*, June 1981, p. 82.
139. A. R. Michaelis, The recovery of science in Germany, *Interdisciplinary Sci. Rev.* **6** (1981) 283.
140. K. E. Zimen, *Bild der Wissenschaft* **5** (1982) 8; G. Obermair, *ibid.*
141. *Phys. Blätter* **37** (1981) 229.
142. *Nature* **291** (1981) 374.
143. *Nature* **293** (1981) 422.
144. A. Herrmann, *Phys. Blätter* **38** (1982) 36.
145. Herrmann had just published a book with the provocative and somewhat misleading title "Wie die Wissenschaft ihre Unschuld verlor" (How science lost its innocence), dedicated mostly to German science in the first half of the 20th century.
146. Private communication, February 24, 1981.
147. Private communication, February 2, 1981.
148. *Allgemeine Jüdische Wochenzeitung*, January 16 and April 24, 1981.
149. Solitary waves were discovered by J. S. Russell in 1834 in the Edinburgh–Glasgow canal.
150. A. E. Glassgold *et al.*, *Annals Phys.* **6** (1959) 1.
151. Initially this was due to the interest in shock waves, which appear above the velocity of sound. Later the interest in high-energy heavy ion reactions was stimulated by the search for quark–gluon plasma.
152. G. N. Fowler, S. Raha, N. Stelte and R. M. Weiner, *Phys. Lett. B* **115** (1982) 286.

153. J. Galin *et al.*, *Phys. Rev. Lett.* **48** (1982) 1787.
154. S. Raha and R. M. Weiner, *Phys. Rev. Lett.* **50** (1983) 407.
155. M. Lefort, *Nucl. Phys. A* **387** (1983) 3c.
156. S. Raha, K. Wehrberger and R. M. Weiner, *Nucl. Phys. A* **433** (1984) 427.
157. E. F. Hefter, S. Raha and R. M. Weiner, *Phys. Rev. C* **32** (1985) 2201.
158. G. N. Fowler, S. Raha and R. M. Weiner, *Phys. Rev. C* **31** (1985) 1515.
159. T. Maxworthy and L. G. Redekopp, *Nature* **260** (1976) 509; *Icarus* **29** (1976) 261.
160. E. Madelung, *Z. Phys.* **40** (1926) 322.
161. Among other things, it has the inconvenience of being non-relativistic.
162. Vorticity is the curl of the fluid velocity; it is a measure of the rate of rotation of the fluid.
163. *Kieler Nachrichten*, January 26, 1983.
164. *Tribüne*, Vol. 33, No. 88, p. 104.
165. Brentano and his family, although non-Jewish, had been emigrants; Hagedorn lived in Switzerland since the 1950s; Rudolf Haag had spent quite a long time in the States; and Brandt had got his PhD in the States.
166. E. Klee, *Die Zeit*, January 27, 2000.
167. H. Markl, *Die Zeit*, February 10, 2000.
168. H. Markl, *Symposium on Biological Sciences and experiments on humans at the institutes of the KWG — The connection with Auschwitz*, Berlin, June 7, 2000, quoted in H. Schuh, *Die Zeit*, June 13, 2001.
169. A. Herrmann, *Die Neue Physik*, München 1979, p. 118.
170. G. N. Fowler and R. M. Weiner, Application of the methods of quantum optics to multihadron production, in *Hadronic Multiparticle Production*, ed. P. Carruthers, World Scientific 1988, p. 481.
171. Quantum physics deals with states — eigenstates — that indicate the probability to find a certain physical property — eigenvalue — in that particular state. Examples of such properties are momentum, coordinate, number of particles, etc. The German word *eigen* means here characteristic.
172. Gauge theories are theories in which the interaction is determined by the symmetry of the system.
173. The correlation function of order *n* defines the transition amplitudes between states containing *n* particles.
174. Superconductivity was discovered in 1912 by Kamerlingh Onnes. The phenomenon of Bose condensation was predicted by Sathyendra Nath Bose and Albert Einstein in 1924 and superfluidity was discovered by Kapitza in 1938. Bose Einstein correlations, as such, were experimentally put in evidence for the first time in the 1950s by the Hanbury–Brown and Twiss intensity interference experiments with stellar light.
175. A characteristic property of interference is that the intensity of the resultant wave (field) is different from the sum of the intensities of the initial components.
176. G. Goldhaber, S. Goldhaber W. Lee and A. Pais, *Phys. Rev.* **120** (1960) 300, reprinted in Ref. 193, p. 2.

177. V. G. Grishin, G. I. Kopylov and M. I. Podgoretski¡i, *Sov. J. Nucl. Phys.* **13** (1971) 638, reprinted in Ref. 193, p. 16.
178. G. N. Fowler and R. M. Weiner, *Phys. Lett. B* **70** (1977) 201.
179. An attempt to disentangle in Bose–Einstein correlations the effects of coherence from final state strong interactions was performed by us in G. N. Fowler, N. Stelte and R. M. Weiner, *Nucl. Phys. A* **319** (1979) 349, reprinted in *Coherent States*, eds. J. R. Klauder and B. Skagerstam, World Scientific 1985, p. 762.
204. G. N. Fowler and R. M. Weiner, *Phys. Rev. D* **17** (1978) 3118, reprinted in Ref. 193, p. 78.
181. M. Gyulassy, S. K. Kaufmann and L. W. Wilson, *Phys. Rev. C* **20** (1979) 2267, reprinted in Ref. 193, p. 86.
182. R. M. Weiner, *Phys. Lett. B* **232** (1989) 278 and *ibid. B* **218** (1990), reprinted in Ref. 193, p. 284.
183. N. Neumeister *et al.*, *Phys. Lett. B* **275** (1992) 186, reprinted in Ref. 193, p. 332.
184. Not surprisingly the correction factor turned out to be $[2/n(n-1)]^{1/2}$.
185. M. Plümer, L. V. Razumov and R. M. Weiner, *Phys. Lett. B* **286** (1992) 335, reprinted in Ref. 193, p. 344.
186. Our fourth LESIP meeting which inaugurated this trend was accordingly entitled *Correlations and Multiparticle Production* (*CAMP*), eds. M. Plümer, S. Raha and R. M. Weiner, World Scientific 1991.
187. I. V. Andreev and R. M. Weiner, *Phys. Lett.* **253** (1991) 416, reprinted in Ref. 193, p. 312.
188. I. V. Andreev, M. Plümer and R. M. Weiner, *Phys. Rev. Lett.* **67** (1991) 3475, reprinted in Ref. 193, p. 326; I. V. Andreev, M. Plümer and R. M. Weiner, *Int. J. Mod. Phys. A* **8** (1993) 4577.
189. L. V. Razumov and R. M. Weiner, *Phys. Lett. B* **348** (1995) 133, reprinted in Ref. 193, p. 452.
190. M. Bowler, *Phys. Lett. B* **276** (1992) 237.
191. R. M. Weiner, *Phys. Rep.* **327** (2000) 249.
192. R. M. Weiner, *Introduction to Bose–Einstein Correlations and Subatomic Interferometry*, John Wiley & Sons, Chichester, New York, 2000.
193. R. M. Weiner (ed.), *Bose–Einstein Correlations in Particle and Nuclear Physics, A Collection of Reprints*, John Wiley & Sons, 1997.
194. The quantum mechanical concept of state used here should be distinguished from the thermodynamical concept of *state of matter*. Thus, e.g. a condensate is from the quantum mechanical point of view a coherent state and at the same time a thermodynamical state of matter.
195. The limitations of this approximation, which is also known under the name "boost-invariance," are discussed in Chap. 29.
196. G. N. Fowler, E. M. Friedlander and R. M. Weiner, *Phys. Lett. B* **104** (1981), 239.
197. G. N. Fowler, E. M. Friedlander, F. W. Pottag, J. Wheeler and G. Wilk, *Phys. Rev. D* **37** (1988) 3127.

198. A. Bialas and R. Peschanski, *Nucl. Phys. B* **273** (1986) 703.
199. P. Carruthers, E. M. Friedlander, C. C. Shih and R. M. Weiner, *Phys. Lett.* **222** (1989) 487; I. V. Andreev, M. Plümer, B. R. Schlei and R. M. Weiner, *Phys. Lett. B* **316** (1993) 583, reprinted in Ref. 193, p. 402.
200. M. Tannenbaum, *Mod. Phys. Lett. A* **9** (1994) 89; *Phys. Lett. B* **347** (1995) 431.
201. G. N. Fowler, E. M. Friedlander, R. M. Weiner and G. Wilk, *Phys. Rev. Lett.* **57** (1986) 2119.
202. I. V. Andreev, M. Plümer, B. R. Schlei and R. M. Weiner, *Phys. Lett. B* **321** (1994) 277, reprinted in Ref. 193, p. 438.
203. G. N. Fowler, E. M. Friedlander, R. M. Weiner and G. Wilk, *Phys. Rev. Lett.* **56** (1986) 14, reprinted in Ref. 193, p. 446.
204. S. L. McCall and E. L. Hahn, *Phys. Rev. Lett.* **18** (1967) 908.
205. Stimulated emission is a quantum phenomenon, which plays an important role in lasers.
206. S. Z. Belenkii and L. D. Landau, *Usp. Fiz. Nauk* **56** (1955) 309.
207. J. Cohen *et al.*, *Lett. Nuovo Cimento* **9** (1974) 337.
208. Phenix White paper in http://www.bnl.gov/bnlweb/pubaf/pr/RHIC-peer.asp, *Nucl. Phys. A* 757(1–2) (2005).
209. R. Dicke, *Phys. Rev.* **93** (1954) 99.
210. G. N. Fowler and R. M. Weiner, *Phys. Rev. Lett.* **55** (1985) 1373, reprinted in Ref. 193, p. 164.
211. G. N. Fowler, E. M. Friedlander and R. M. Weiner, *Phys. Lett. B* **116** (1982) 203.
212. W. Schleich and J. A. Wheeler, *Nature* **326** (1987) 574.
213. A. Vourdas and R. M. Weiner, *Phys. Rev. A* **36** (1987) 5866.
214. A. Vourdas and R. M. Weiner, *Phys. Rev. D* **38** (1988) 2209, reprinted in Ref. 193, p. 244.
215. I. V. Andreev and R. M. Weiner, *Phys. Lett. B* **373** (1996) 159.
216. The situation appears to be quite different in the former Central Europe satellite countries, which have joined the European Union.
217. A. Abragam, *De la Physique d'avant toute chose*, Odile Jacob 1987.
218. Cf. e. g. the case of the conscientious objector Michael Lepscher executed by the Nazis; his family had to pay a 300 Reichsmark fine for the death sentence, a fee of 158. 18 Reichsmark for the execution and a fee of 1.5 Reichsmark per detention day; *Die Zeit*, May 24, 2006.
219. "Kuhfladenroulette" by U. Schnabel, *Die Zeit*, December 30, 2004.
220. I. Rioufol, *Le Figaro*, March 31, 2006.
221. See e. g. *Die Zeit*, May 20, 1994.
222. R. M. Weiner, *Int. J. Mod. Phys.* **15** (2006) 37.
223. J. C. Collins and M. J. Perry, *Phys. Rev. Lett.* **34** (1975) 1353.
224. E. V. Shuryak, *Phys. Lett. B* **78** (1978) 150.
225. A. Mann and R. M. Weiner, *Lett. Nuovo Cimento* **2** (1971) 248.
226. A phase transition implies the equality of the free energies of the two phases.
227. M. Plümer, S. Raha and R. M. Weiner, *Nucl. Phys. A* **418** (1984) 549c.

228. G. Baym, in *Quark Matter Formation and Heavy Ion Collisions*, eds. M. Jacob and H. Satz, World Scientific 1982.
229. G. N. Fowler, S. Raha and R. M. Weiner, *Z. Phys. C* **9** (1981) 271.
230. http://www.bnl.gov/bnlweb/pubaf/pr/RHIC-peer.asp; *Nucl. Phys. A* 757(1–2) (2005).
231. E. V. Shuryak, *Prog. Part. Nucl. Phys.* **53** (2004) 273; M. Gyulassy and L. McLerran, *Nucl. Phys. A* **750** (2005) 30.
232. A. Chodos, R. L. Jaffe, K. Johnson, C. B. Thorn and V. Weisskopf, *Phys. Rev. D* **9** (1974) 3471.
233. C. Nosaka, E. Honda and S. Muroya, *Eur. Phys. J. C* **17** (2000) 663; T. Hirano, K. Morita, S. Muroya and C. Nonaka, nucl-th/0110009.
234. M. Plümer, S. Raha and R. M. Weiner, *Nucl. Phys. A* **418** (1984) 549c; M. Plümer, S. Raha and R. M. Weiner, *Phys. Lett. B* **139** (1984).
235. For a review of the Landau model cf. E. L. Feinberg, Relativistic heavy ion physics, *Int. Rev. Nucl. Phys.* **6** (1991) 341 and the round table discussion "The Landau hydrodynamical model" in *Local Equilibrium in Strong Interactions Physics* (*LESIP I*), eds. D. K. Scott and R. M. Weiner, World Scientific 1985.
236. K. Wehrberger and R. M. Weiner, *Phys. Rev. D* **31** (1985) 222. As far as I can gather this is the first application of the hydrodynamical model in which a realistic EOS with a variable velocity of sound was used (one of the variants considered included also the influence of confinement on the QGP EOS). This procedure has now become standard in heavy ion physics.
237. This is presumably due to the fact that the entropy of the system is fixed before the (adiabatic) expansion of the system starts.
238. E. M. Friedlander and R. M. Weiner, *Phys. Rev. Lett.* **43** (1979) 15.
239. I. M. Khalatnikov, *Zh. Eksp. Theor. Fiz.* **26** (1954) 529.
240. O. V. Zhirov and E. V. Shuryak, *Yad. Fiz.* **21** (1975) 861 [*Sov. J. Nucl. Phys.* **21** (1975) 443].
241. G. N. Fowler, E. M. Friedlander, R. M. Weiner and G. Wilk, *Phys. Rev. Lett.* **57** (1986) 2119.
242. This universality was observed much earlier in purely hadronic reactions, when comparing nucleon–nucleon with meson–nucleon reactions. In the case of e^+–e^- reactions it also answers the objection of some critics of the application of statistical methods to these reactions, who argue that here the produced particles had no chance to interact and therefore equilibrium could not be reached: The point is that equilibrium is presumably reached at the QGP level.
243. F. Halzen and H. C Liu, *Phys. Rev. D* **25** (1982) 1842.
244. S. Raha, R. M. Weiner and J. Wheeler, *Phys. Rev. Lett.* **53** (1984) 138.
245. K. B. Bhalla *et al.*, *Nucl. Phys. A* **367** (1981) 446.
246. H. G. Baumgardt, E. M. Friedlander and E. Schopper, *J. Phys. G* 7 (1981) L175.
247. D. Ghosh *et al.*, *J. Phys. G: Nucl. Part. Phys.* **18** (1992) 935, and references therein.

248. U. Ornik and R. M. Weiner, *Phys. Rev. D* **36** (1987) 1263.
249. U. Ornik, F. W. Pottag and R. M. Weiner, *Phys. Rev. Lett.* **63** (1989) 2641.
250. N. Masuda and R. M. Weiner, *Phys. Rev. D* **18** (1978) 1515; *ibid.* 1542.
251. U. Ornik, F. W. Pottag and R. M. Weiner, *Phys. Rev. Lett.* **63** (1989) 2641; B. R. Schlei *et al.*, *Phys. Lett. B* **293** (1992) 275; J. Bolz *et al.*, *Phys. Rev. C* **46** (1992) 2047; *Phys. Lett. B* **300** (1993) 404; *Phys. Rev. D* **47** (1993) 3860; U. Ornik *et al.*, *Phys. Rev. C* **54** (1996) 1381.
252. These indications include suppression of bound states of charm–anticharm and jet quenching, which are no longer considered sufficient conditions for QGP.
253. B. R. Schlei *et al.*, *Phys. Lett. B* **293** (1992) 275.
254. B. Gelman, E. Shuryak and I. Zahed, nucl-th/0410067.
255. CERN press release: http://info.web.CERN.ch/Press/PressReleases/Releases 2000/PR01.00EQuark GluonMatter.html
256. http://www.bnl.gov/bnlweb/pubaf/pr/RHIC-peer.asp; *Nucl. Phys. A* **757**(1–2) (2005).
257. *Das Miniatom-Projekt*, LiteraturWissenschaft. de (TransMIT) 2006.
258. Alsos was a secret mission that followed the advancing Allied forces in Europe to determine the progress of Germany's atomic bomb project.
259. R. M. Weiner, *Lett. Nuovo Cimento* 8 (1973) 281.

Index